T0184909

Communications in Computer and Information Science 1261

Commenced Publication in 2007
Founding and Former Series Editors:
Simone Diniz Junqueira Barbosa, Phoebe Chen, Alfredo Cuzzocrea,
Xiaoyong Du, Orhun Kara, Ting Liu, Krishna M. Sivalingam,
Dominik Ślęzak, Takashi Washio, Xiaokang Yang, and Junsong Yuan

More information about this series at http://www.springer.com/series/7899

Cheng He · Mengling Feng ·
Patrick P. C. Lee · Pinghui Wang ·
Shujie Han · Yi Liu (Eds.)

Large-Scale Disk Failure Prediction

PAKDD 2020 Competition and Workshop, AI Ops 2020
February 7 – May 15, 2020
Revised Selected Papers

 Springer

Editors
Cheng He
Alibaba Group (China)
Hangzhou, China

Mengling Feng 🆔
National University of Singapore
Singapore, Singapore

Patrick P. C. Lee 🆔
Chinese University of Hong Kong
Hong Kong, China

Pinghui Wang
Xi'an Jiaotong University
Xi'an, China

Shujie Han 🆔
Chinese University of Hong Kong
Hong Kong, China

Yi Liu
Alibaba Group (China)
Hangzhou, China

ISSN 1865-0929   ISSN 1865-0937   (electronic)
Communications in Computer and Information Science
ISBN 978-981-15-7748-2   ISBN 978-981-15-7749-9   (eBook)
https://doi.org/10.1007/978-981-15-7749-9

This Springer imprint is published by the registered company Springer Nature Singapore Pte Ltd.
The registered company address is: 152 Beach Road, #21-01/04 Gateway East, Singapore 189721, Singapore

Preface

This volume contains the top papers from the PAKDD 2020 Alibaba AI Ops Competition on Large-Scale Disk Failure Prediction. The competition was conducted between February 7 to May 15, 2020, on the Alibaba Cloud Tianchi Platform (https://tianchi.aliyun.com/competition/entrance/231775/introduction).

The competition aims to develop machine learning models to accurately predict disk failures in the large-scale data centers. Nowadays, the number of hard disk drive (HDD) and solid-state drive (SSD) has reached millions in large data centers, where disk failures account for the largest proportion of all failures. The frequent occurrence of disk failures will affect the stability and reliability of the server and even the entire IT infrastructure. Therefore, it is desirable for large-scale data centers to have an effective tool to predict disk failures to allow early prevention and timely maintenance. However, solving this problem is not a trivial task owing to a number of data-related challenges, such as high level of noises in the data, extremely imbalanced class distribution, and time-varying feature characteristics.

The Pacific-Asia Conference on Knowledge Discovery and Data Mining (PAKDD) is one of the longest established and leading international conferences in the areas of data mining and knowledge discovery. The event was co-organized by both the PAKDD 2020 committee and the Alibaba Cloud team. It provides an international forum for researchers and industry practitioners to share their new ideas, original research results, and practical development experiences from all KDD related areas, including data mining, data warehousing, machine learning, artificial intelligence, databases, statistics, knowledge engineering, visualization, decision-making systems, and the emerging applications. Alibaba Cloud, also known as Aliyun, is a Chinese cloud computing company, a subsidiary of Alibaba Group. Alibaba Cloud provides cloud computing services to online businesses and Alibaba's own e-commerce ecosystem. Alibaba Cloud's international operations are registered and headquartered in Singapore. Tianchi is a platform hosted on Alibaba Cloud to support data competitions around the world. During PAKDD 2020, we organized a dedicated workshop to feature the best performing teams of the competition. Due to the COVID-19 pandemic, the workshop was hosted online.

We attracted 1,176 teams in total for the competition, and we selected the winners in three phases. For the qualification phase, we selected the top 150 teams; for the semi-finals phase, we selected the top 50 teams; and 12 teams with top 10 best scores (due to some ties) entered the final. All teams who entered the semi-finals were invited to submit their manuscript. In the end, 11 papers were published in this proceeding. All the accepted papers were peer reviewed by two qualified reviewers chosen from our Scientific Committee based on their qualifications and experience in a single-blind manner.

The proceedings editors wish to thank the dedicated Scientific Committee members and all the other reviewers for their contributions. We also thank Springer for their trust and for publishing the proceedings for the PAKDD 2020 Alibaba AI Ops Competition on Large-Scale Disk Failure Prediction.

June 2020

Cheng He
Mengling Feng
Patrick. P. C. Lee
Pinghui Wang
Yi Liu

Organization

Scientific Committee

Cheng He	Alibaba Group, China
Jiongyu Liu	Alibaba Group, China
Pinan Chen	Alibaba Group, China
Fan Xu	Alibaba Group, China
Yi Qiang	Alibaba Group, China
Mian Wang	Alibaba Group, China
Fan Xu	Alibaba Group, China
Yi Liu	Alibaba Group, China
Tao Huang	Alibaba Group, China
Rui Li	Alibaba Group, China
Shujie Han	Chinese University of Hong Kong, China
Fei Wang	Weill Cornell Medicine, USA
Dilruk Perera	National University of Singapore, Singapore
Mengling Feng	National University of Singapore, Singapore
Hady W. Lauw	Singapore Management University, Singapore

Organizing Committee

Xiaoxue Zhao	Alibaba Group, China
Patrick. P. C. Lee	Chinese University of Hong Kong, China
Pinghui Wang	Xi'an Jiaotong University, China
Shiwen Wang	Alibaba Group, China
Zengyi Lu	Alibaba Group, China
Cheng He	Alibaba Group, China
Jiongyu Liu	Alibaba Group, China
Yi Liu	Alibaba Group, China
Fan Xu	Alibaba Group, China
Tao Huang	Alibaba Group, China

Local Committee

Mengling Feng	National University of Singapore, Singapore
Hady W. Lauw	Singapore Management University, Singapore

Contents

An Introduction to PAKDD CUP 2020 Dataset

Yi Liu[1(✉)], Shujie Han[2], Cheng He[1], Jiongzhou Liu[1], Fan Xu[1], Tao Huang[1], and Patrick P. C. Lee[2]

[1] Alibaba Group, Hangzhou, China
978355734@qq.com
[2] The Chinese University of Hong Kong, Shatin, Hong Kong

Abstract. With the rapid development of cloud services, disk storage has played an important role in large-scale production cloud systems. Predicting imminent disk failures is critical for maintaining data reliability. Our vision is that it is important for researchers to contribute to the development of new techniques for accurate and robust disk failure prediction. If researchers can discover any reasonable approaches for disk failure prediction in large-scale cloud systems, all IT and big data companies can benefit from such approaches to further enhance the robustness of the production cloud systems. With this vision in mind, we have published an open labeled dataset that spans a period of 18 months with a total of 220,000 hard drives collected from Alibaba Cloud. Our dataset is among the largest released in the community in terms of its scale and duration. To better understand our dataset, we present our dataset generation process and conduct a preliminary analysis on the characteristics of our dataset. Our open dataset has been adopted in the PAKDD2020 Alibaba AI Ops Competition, in which contestants proposed new disk failure prediction algorithms through the analysis and evaluation of the dataset.

Keywords: Hard disk drive · PAKDD2020 · Alibaba Cloud

1 Introduction

The rapid development of cloud services motivates the need of big data storage infrastructures for managing an ever-increasing amount of data. Today's cloud providers often deploy production data centers that are equipped with millions of hard disk drives spanning across the globe [26]. With such large-scale deployment of data centers, disk failures are commonplace. Field studies show that disk failures account for the largest proportion of hardware failures in cloud data centers, and the annual failure rate of hard disk drives is in the range of 2–4% and even up to 13% in some observed systems [22]. This implies that administrators need to handle hundreds of disk failures/replacements in production data centers

Jointly organized by PAKDD 2020, Alibaba Cloud and Alibaba Tianchi Platform.

© Springer Nature Singapore Pte Ltd. 2020
C. He et al. (Eds.): AI Ops 2020, CCIS 1261, pp. 1–11, 2020.
https://doi.org/10.1007/978-981-15-7749-9_1

on a daily basis. The frequent occurrences of disk failures challenge both data availability and durability. If such disk failures cannot be properly resolved, this will pose a negative impact on business service-level agreements.

To maintain data availability and durability guarantees, it is critical for administrators to proactively predict imminent disk failures before they actually happen. *Disk failure prediction* has been an important topic for IT or big data company. In the past 15 years, a variety of valuable studies propose various data-driven techniques (e.g., machine learning models) on boosting the accuracy of disk failure prediction [5,7,9,10,13,16,18,24]. Such studies build on various types of available datasets for their designs. The most commonly used data type is based on SMART (Self-Monitoring, Analysis and Reporting Technology) [4], which is widely used for monitoring the healthy status of hard disk drives. However, SMART is not completely standardized. Indeed, the collected set of SMART attributes and the SMART implementation details vary across different hard drive vendors [6]. Thus, some other types of data are also adopted, such as system-level signals (e.g., windows events, file system operation error, unexpected telemetry loss, etc.) [26], as well as performance data (at both disk and server levels) and location attributes (site, room, rack, and server) [17].

1.1   Challenges of Disk Failure Prediction

The accuracy of a disk failure prediction algorithm heavily depends on the input dataset, yet designing a highly accurate disk failure prediction is often subject to the following challenges due to the inherent characteristics of the input dataset itself:

- **Data noise**. The data noise is mainly attributed to the labeling noise and the sampling noise. Typically, expert rules are used to label the disk failures according to the prior experience. Thus, expert rules are not able to cover the unknown failure types, thereby leading to false negatives. Also, expert rules are simple by nature, and hence they are not capable of dealing with complex failure types, thereby leading to false positives. Furthermore, unexpected accidents may interrupt in the data collection, thereby introducing the missing values or sampling noise. Han *et al.* [13] propose a robust data preprocessing method to deal with the data noise issue for the disk failure prediction problem under the imperfect data quality.
- **Data imbalance**. In disk failure prediction, the proportion of healthy disks is always much larger than that of failed disks, leading to data imbalance. To mitigate the impact of data imbalance problem, prior studies often utilize the down-sampling method to balance the ratio between the positives and negatives [6,7,18]. As mentioned in [26], rebalancing methods help raise the recall, but introduce a large number of false positives at the same time, thereby dramatically decreasing the precision. Meanwhile, Xu *et al.* [26] indicate that the ranking method can mitigate the data imbalance problem because it is insensitive to the class imbalance. Furthermore, Lu *et al.* [17] use the F-measure [14] and MCC [8] as their evaluation metrics to cope with the data imbalance

problem. Han *et al.* [13] propose to mark the pre-failure samples of failed disks as positive by automated pre-failure backtracking.

- **Time-varying features**. Due to the aging of disks and the additions/removals of disks in production, the statistical patterns of disk logs are varying over time [12]. Han *et al.* [12] use the two-sample Kolmogorov-Smirnov (KS) test [19] to measure the change of the distributions of SMART attributes and present a general stream mining framework for disk failure prediction with concept-drift adaptation.

1.2 Existing Open Datasets

Prior studies on disk failure prediction are based on either proprietary datasets or open datasets. Examples of existing open datasets for disk failure prediction include the following.

- **CMRR** [21]. The SMART dataset was published by the Center for Memory and Recording Research (CMRR). It covers 369 hard disks of a single drive model. Each disk is labeled as either good or failed, with 178 good disks and 191 failed disks.
- **Baidu** [2]. The SMART dataset was collected from an enterprise-class disk model of Seagate ST31000524NS at Baidu. It covers 23,395 disks. Each disk is labeled as either good or failed, with only 433 failed disks and 22,962 healthy disks. The dataset also reports 14 SMART attributes, which were collected on hourly basis and normalized to the interval from −1 to 1 (inclusively).
- **Backblaze** [1]. The SMART dataset was collected by Backblaze and have been extensively used for the evaluation in the literature [7,12,18,24]. As of September 30, 2019, the dataset covers 112,864 disks among 13 disk models from three vendors, spanning a period from April 2013 to September 2019.
- **WSU** [17]. The dataset was collected at an anonymized enterprise by the research team at Wayne State University (WSU). It covers 380,000 hard disks over a period of two months across 64 sites. The dataset reports not only the SMART attributes, but also the performance (disk-level and server-level) data and location attributes of disks (site, room, rack, and server).

The maximum number of disks from a single disk model in the Backblaze, Baidu, and CMRR is less than 50,000, which is generally smaller than that of large-scale production systems. The WSU dataset, while covering more than 380,000 disks, only contains two months of data. We believe that an open dataset with a larger scale of disks and a long duration of operations will be beneficial for the research community to develop disk failure prediction methods.

1.3 Our Contributions

In this work, we introduce an open SMART dataset collected at Alibaba Cloud. It covers a total of 220,000 hard disks that are deployed at data centers, spanning a period of 18 months. Our dataset is among the largest released in the community, in terms of its scale (compared to CMRR, Baidu, and Backblaze) and duration (compared to WSU [17]). It was also adopted in the PAKDD2020 Alibaba AI Ops Competition for contestants to design new disk failure prediction algorithms. We hope that more researchers can benefit from our open dataset in enhancing the work of disk failure prediction. Our dataset is available at: https://github.com/alibaba-edu/dcbrain.

In the following, we first describe the generation process of our open dataset (Sect. 2). We next present a preliminary analysis on the characteristics of the dataset, including annualized failure rates (AFR) statistics, data missing, and the SMART statistics (Sect. 3). Finally, we review the related work (Sect. 4) and conclude the paper (Sect. 5).

2   Dataset Generation

In this section, we introduce the generation process of our open dataset, including sampling and data anonymization.

2.1   Sampling

We formulate the sampling process as an optimization problem whose objective is to minimize the distribution differences between the original dataset and the sampled dataset. Each disk has its unique identifier and emits the SMART attributes over a time series. We denote the sets of identifiers for healthy and failed disks by $\mathcal{O}_h$ and $\mathcal{O}_f$, respectively. We denote the SMART attributes of the original dataset by a vector $\mathbf{x}_O$. Note that $\mathbf{x}_O$ is a subset of the *whole* collection of disks. Specifically, $\mathbf{x}_O$ consists of the time-series samples of the last 30 days with the six most important SMART attributes (i.e., SMART-5, SMART-187, SMART-188, SMART-193, SMART-197, and SMART-198). Note that there exists data missing of the SMART attributes on some days (see Sect. 3.2). We use the forward filling method to fill the missing values.

We adopt *stratified sampling* for healthy and failed disks to keep the ratio of healthy to failed disks. We select the healthy and failed disks randomly from $\mathcal{O}_h$ and $\mathcal{O}_f$, respectively, with the sampling ratio r. The sets of sampled disks are denoted by $\mathcal{S}_h$ for healthy disks and $\mathcal{S}_f$ for failed disks. The distributions of sampled dataset $\mathbf{x}_S$ consist of the time-series samples of the last 30 days healthy disks $\mathcal{S}_h$ and the last 30 days before failure occurrences for failed disks $\mathcal{S}_f$.

We use the maximum mean discrepancy (MMD) [10] to measure the distribution differences of the SMART attributes between $\mathbf{x}_O$ and $\mathbf{x}_S$. We denote the MMD by ϵ. If ϵ is closer to zero, it means that the two distributions are more similar.

Algorithm 1 shows the pseudo-code of the whole workflow for sampling. The MAIN procedure takes the inputs of $\mathcal{O}_h$, $\mathcal{O}_f$, $\mathbf{x}_O$, and r. It performs initialization (Lines 2–5) and executes over a number of iterations n, where n is a tunable parameter. In each iteration, it randomly samples the identifiers for healthy and failed disks from $\mathcal{O}_h$ and $\mathcal{O}_f$, and keeps them into $\mathcal{S}_h$ and $\mathcal{S}_f$, respectively

---

Algorithm 1. Framework of data sampling.

1: **procedure** MAIN($\mathcal{O}_h$, $\mathcal{O}_f$, $\mathbf{x}_O$, r)
2:      Initialize ϵ_{min} = Infinite
3:      Initialize $\mathcal{S}_{min}$ = empty set
4:      Initialize $\mathcal{S}_h$ = empty set
5:      Initialize $\mathcal{S}_f$ = empty set
6:      **for** $i = 1$ to n **do**
7:          $\mathcal{S}_h$ = Choose disk identifiers randomly from $\mathcal{O}_h$ with r
8:          $\mathcal{S}_f$ = Choose disk identifiers randomly from $\mathcal{O}_f$ with r
9:          Update $\mathbf{x}_S$ with the time-series samples of the last 30 days for $\mathcal{S}_h$ and $\mathcal{S}_f$
10:          Compute ϵ between $\mathbf{x}_O$ and $\mathbf{x}_S$
11:          **if** $\epsilon < \epsilon_{min}$ **then**
12:              $\mathcal{S}_{min} = \mathcal{S}_h + \mathcal{S}_f$
13:          **end if**
14:      **end for**
15:      **return** All time-series samples of $\mathcal{S}_{min}$
16: **end procedure**

(Lines 7–8). It updates $\mathbf{x}_S$ with the time-series samples of the last 30 days for $\mathcal{S}_h$ and $\mathcal{S}_f$ (Line 9). Then it computes the MMD between $\mathbf{x}_O$ and $\mathbf{x}_S$ (Line 10). If the current ϵ is less than ϵ_{min}, it updates the set $\mathcal{S}_{min}$ with the union set of $\mathcal{S}_h$ and $\mathcal{S}_f$ (Lines 11–13). It returns the time-series samples of $\mathcal{S}_{min}$ (Line 15).

2.2   Data Anonymization

Due to privacy concerns, we anonymize the sensitive fields in the dataset. More concretely, we use "manufacturer" "k" to represent each disk model, where "manufacturer" corresponds to a letter ("A"), and "k" (1 to 2) corresponds to the k-th numerous model; for example, "A1" represents the most numerous disk model of vendor A. Also, we sort the disks by the serial numbers and reset the disk identifiers as the order of disks.

3   Dataset Analysis

We generate the open dataset over a time period from July, 2017 to December, 2018. Table 1 shows the overview of the two disk models, both of which are SATA hard disk drives (HDDs). The total counts of both A1 and A2 are over 100,000 each, which is larger than any single disk model reported in the Backblaze dataset. Compared to the dataset in [17], the time span of our open dataset is over a period of 18 months, which is more beneficial for researchers to study the temporal change of failure patterns in disk failure prediction issues.

Table 1. Overview of disk models A1 and A2 in our open dataset.

Type	Model	Disk count	# failures
HDD_SATA	A1	106,453	1,243
HDD_SATA	A2	102,779	1,162

3.1   Failure Rates

We first estimate the annualized failure rates (AFRs) of both disk models A1 and A2. Specifically, we define the AFR as the ratio between the number of failed disks reported in our trouble tickets during the one-year span of our dataset and the total number of disks. Table 2 shows that the AFRs of A1 and A2 are 0.90% and 1.01%, respectively.

Table 2. AFRs for disk models A1 and A2.

Type	Model	AFR (%)
HDD_SATA	A1	0.90
HDD_SATA	A2	1.01

To study the failure rates in a more fine-grained manner, we further compute the *monthly failure rates (MFRs)* of disks in our dataset from July, 2017 to December, 2018. Similarly, we define the MFR as the ratio between the number of failed disks reported in our trouble tickets during a one-month span of our dataset and the total number of disks. Figure 1 shows that the MFRs of both A1 and A2 have increasing trends.

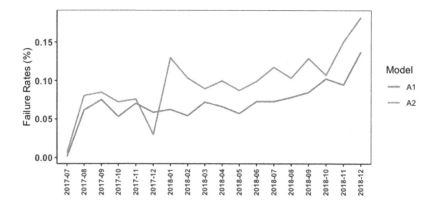

Fig. 1. MFRs for disk models A1 and A2.

We try to explain the phenomenon of the increasing trends of the MFRs by the *workload* of HDDs. We define the workload as the total number of 512-byte sectors written and read on a daily basis, which is computed from the two SMART attributes, i.e., SMART-241 ("Total LBAs Written") and SMART-242 ("Total LBAs Read"). Figure 2 shows the average workload in each month for both disk models A1 and A2. The figure suggests that the increasing trends of the MFRs of both A1 and A2 may be attributed to the increasing workload.

We finally study the failure rates across different days of a week. Figure 3 shows that the failure rates of both A1 and A2 on weekends are lower than those on weekdays. We explain this phenomenon by computing the average workloads over all weekdays and weekends. Table 3 shows that for both A1 and A2, the average workload on weekdays is heavier than that on weekends, which implies that the heavier workload of disks may cause the higher failure rates.

3.2   Data Missing

Due to the complexity of large-scale production systems, some unexpected accidents may interrupt the data collection and lead to data missing. In our dataset, data missing exists in both failed and healthy disks. To better describe the severity of data missing in the competition dataset, we introduce the data missing ratio (DMR), defined by the ratio between the actual missing days and the expected occurrence days if no data missing occurs. For healthy disks, the expected occurrence days are from when the disks first appear in the dataset until the end day of the collection time; for failed disks, the expected occurrence days are from when the disks first appear in the dataset until the reported date of the trouble ticket.

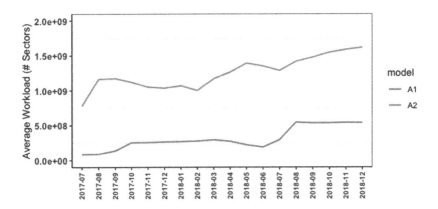

Fig. 2. Average workload in each month for disk models A1 and A2.

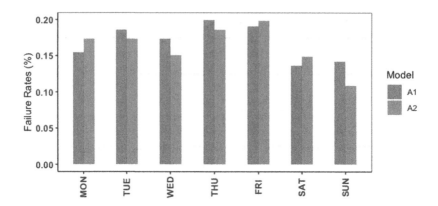

Fig. 3. Failure rates in different days of a week for disk models A1 and A2.

Table 3. Average workload over all weekdays and weekends.

Model	Average workload (# sectors)	
	Weekdays	Weekends
A1	$3.5e + 08$	$3.1e + 08$
A2	$1.34e + 09$	$1.31e + 09$

Table 4. DMRs for disk models A1 and A2 for failed and healthy disks.

Model	DMR of failed disks	DMR of healthy disks
A1	11.9%	6.6%
A2	6.9%	5.6%

Table 5. Overview of the collected SMART attributes.

ID	SMART attribute name	A1	A2	ID	SMART attribute name	A1	A2
1	Raw read error rate	✓	✓	191	G-sense error rate	✓	✓
3	Spin up time	✓	✓	192	Power-off retract count	✓	✓
4	Start stop count	✓	✓	193	Load cycle count	✓	✓
5	Reallocated sector count	✓	✓	194	Temperature celsius	✓	✓
7	Seek error rate	✓	✓	195	Hardware ECC recovered	✓	✓
9	Power on hours	✓	✓	196	Reallocated event count	✓	
10	Spin retry count	✓	✓	197	Current pending sector	✓	✓
12	Power cycle count	✓	✓	198	Offline uncorrectable error	✓	✓
184	End-to-end error	✓	✓	199	UDMA CRC error count	✓	✓
187	Reported uncorrectable error	✓	✓	240	Head flying hours	✓	✓
188	Command timeout	✓	✓	241	Total LBAs written	✓	✓
189	High fly writes	✓	✓	242	Total LBAs read	✓	✓
190	Airflow temperature celsius	✓	✓				

Table 4 shows that the DMRs for failed disks are 8.1% for A1 and 7.5% for A2, while the DMRs of healthy disks are 6.6% for A1 and 5.6% for A2. The data missing issue in failed disks is be more severe than in healthy disks.

3.3 Statistics of the SMART Attributes

Finally, we analyze the SMART attributes in our dataset. Table 5 shows the collected SMART attributes in our dataset for A1 and A2. A1 has 25 SMART attributes, while A2 has 24 SMART attributes. Each SMART attribute has both the raw and normalized values.

We next analyze the correlation between the SMART attributes and the failures in our dataset. We compute Spearman's Rank Correlation Coefficient (SRCC) between each of the SMART attributes and disk failures. SRCC ranges from −1 to +1 and its absolute value is closer to 1 implies that the two variables are more correlated, while 0 means no correlation.

Table 6 shows the three most failure-correlated SMART attributes for disk model A1 and A2 with the largest absolute SRCC values. We can see that "reallocated sector count", "current pending sector", and "reported uncorrectable error" are the three most failure-correlated SMART attributes of A1 and A2.

Table 6. Spearman's Rank Correlation Coefficients between the SMART attributes and disk failures for disk models A1 and A2.

Model	Rank 1	Rank 2	Rank 3
A1	SMART-197: 0.37	SMART-187: 0.23	SMART-5: 0.21
A2	SMART-5: 0.29	SMART-197: 0.26	SMART-187: 0.24

4 Related Work

Most of previous work with disk failure prediction focus on the HDDs, which show that highly accurate disk failure prediction can be achievable using classical statistical techniques and machine learning models, such as rank-sum tests [15], Bayesian classifiers [11], rule-based learning [5], back-propagation neural networks [27], regularized greedy forests [7], random forests [18], online random forests [24], and stream-based data mining [12]. Most of the studies (e.g., [7,12,18,24] use the Backblaze dataset for their evaluation.

There are some research studies from industry, such as Alibaba Cloud [25], Facebook [20], Google [23], focusing on the solid-state drives (SSDs). However, they do not release the datasets for their SSD failure prediction.

From the algorithm competition, Huawei [3] holds a hackathon contest about HDDs failure prediction. In this competition, it covers 550,000 samples of HDDs, including total 16,348 healthy disks and 1,160 failed disks. Contestants can acquire the datasets in Huawei's NAIVE platform during the competition.

5   Conclusion

In this paper, we present an open dataset covering a period of 18 months with a total of 220,000 hard drives, collecting from Alibaba Cloud. Our open dataset is applied into the PAKDD2020 Alibaba AI Ops Competition. We hope that more researchers can participate in solving the disk failure prediction problem based on our published dataset. In future work, we plan to update our open dataset every half a year, and evaluate the feasibility of publishing more data types in addition to SMART attributes.

References

1. Backblaze datasets. https://www.backblaze.com/b2/hard-drive-test-data.html
2. Baidu datasets. https://pan.baidu.com/share/link?shareid=189977&uk=4278294944#list/path=%2FS.M.A.R.T.dataset
3. Huawei datasets. https://www.hwtelcloud.com/competitions/hdc-2020-competition
4. Wiki on S.M.A.R.T. https://en.wikipedia.org/wiki/S.M.A.R.T
5. Agarwal, V., Bhattacharyya, C., Niranjan, T., Susarla, S.: Discovering rules from disk events for predicting hard drive failures. In: Proceedings of IEEE ICMLA (2009)
6. Aussel, N., Jaulin, S., Gandon, G., Petetin, Y., Fazli, E., Chabridon, S.: Predictive models of hard drive failures based on operational data. In: Proceedings of IEEE ICMLA (2017)
7. Botezatu, M.M., Giurgiu, I., Bogojeska, J., Wiesmann, D.: Predicting disk replacement towards reliable data centers. In: Proceedings of ACM SIGKDD (2016)
8. Boughorbel, S., Jarray, F., El-Anbari, M.: Optimal classifier for imbalanced data using matthews correlation coefficient metric. PLoS ONE **12**(6) (2017)
9. Cidon, A., Escriva, R., Katti, S., Rosenblum, M., Sirer, E.G.: Tiered replication: a cost-effective alternative to full cluster geo-replication. In: Proceedings of USENIX ATC (2015)
10. Gretton, A., Borgwardt, K.M., Rasch, M.J., Schölkopf, B., Smola, A.: A kernel two-sample test. J. Mach. Learn. Res. **13**, 723–773 (2012)
11. Hamerly, G., Elkan, C., et al.: Bayesian approaches to failure prediction for disk drives. In: Proceedings of ACM ICML (2001)
12. Han, S., Lee, P.P.C., Shen, Z., He, C., Liu, Y., Huang, T.: Toward adaptive disk failure prediction via stream mining. In: Proceedings of IEEE ICDCS (2020)
13. Han, S., et al.: Robust Data Preprocessing for Machine-Learning-Based Disk Failure Prediction in Cloud Production Environments. arXiv preprint arXiv:1912.09722 (2019)
14. Hripcsak, G., Rothschild, A.S.: Agreement, the f-measure, and reliability in information retrieval. J. Am. Med. Inform. Assoc. **12**(3), 296–298 (2005)
15. Hughes, G.F., Murray, J.F., Kreutz-Delgado, K., Elkan, C.: Improved disk-drive failure warnings. IEEE Trans. Reliab. **51**(3), 350–357 (2002)
16. Li, J., et al.: Hard drive failure prediction using classification and regression trees. In: Proceedings of IEEE/IFIP DSN (2014)
17. Lu, S.L., Luo, B., Patel, T., Yao, Y., Tiwari, D., Shi, W.: Making disk failure predictions SMARTer! In: Proceedings of USENIX FAST (2020)

18. Mahdisoltani, F., Stefanovici, I., Schroeder, B.: Proactive error prediction to improve storage system reliability. In: Proceedings of USENIX ATC (2017)
19. Massey, F.J.: The Kolmogorov-Smirnov test for goodness of fit. J. Am. Stat. Assoc. **46**(253), 68–78 (1951)
20. Meza, J., Wu, Q., Kumar, S., Mutlu, O.: A large-scale study of flash memory failures in the field. In: Proceedings of ACM SIGMETRICS (2015)
21. Murray, J.F., Hughes, G.F., Kreutz-Delgado, K.: Machine learning methods for predicting failures in hard drives: a multiple-instance application. J. Mach. Learn. Res. **6**(May), 783–816 (2005)
22. Schroeder, B., Gibson, G.A.: Disk failures in the real world: what does an MTTF of 1,000,000 hours mean to you? In: Proceedings of USENIX FAST (2007)
23. Schroeder, B., Lagisetty, R., Merchant, A.: Flash reliability in production: the expected and the unexpected. In: Proceedings of USENIX FAST (2016)
24. Xiao, J., Xiong, Z., Wu, S., Yi, Y., Jin, H., Hu, K.: Disk failure prediction in data centers via online learning. In: Proceedings of ACM ICPP (2018)
25. Xu, E., Zheng, M., Qin, F., Xu, Y., Wu, J.: Lessons and actions: what we learned from 10K SSD-related storage system failures. In: Proceedings of USENIX ATC (2019)
26. Xu, Y., et al.: Improving service availability of cloud systems by predicting disk error. In: Proceedings of USENIX ATC (2018)
27. Zhu, B., Wang, G., Liu, X., Hu, D., Lin, S., Ma, J.: Proactive drive failure prediction for large scale storage systems. In: Proceedings of IEEE MSST (2013)

PAKDD 2020 Alibaba AIOps Competition - Large-Scale Disk Failure Prediction: Third Place Team

Bo Zhou[✉]

CNCERT, Beijing, China
zhoubo@cert.org.cn

Abstract. This paper describes our submission to the PAKDD 2020 Alibaba AIOps Competition: Large-scale Disk Failure Prediction. Our approach is based on LightGBM classifier with focal loss objective function. The method ranks third with a F1-score of 0.4047 in the final competition season, while the winning F1-score is 0.4903.

Keywords: Binary-classification · Disk failure prediction · Focal loss function

1 Problem Description

The goal of Large-scale Disk Failure Prediction Competition is to predict whether a disk will suffer from a imminent failure within the next 30 days. This task is of critical importance to mitigate the risk of data loss, recovery cost and lower reliability in modern data centers. We interpreted this problem as a supervised binary-classification problem, where the label is 1 if a disk is going to crash within the following 30 days and 0 in other cases. SMART (Self-Monitoring, Analysis and Reporting Technology) features were opted as train dataset, as they were supposed to reveal the defect information and gradual degradation of the underlying disks.

The complexities and challenges in the competition can be summarized as: (1) The amount of data was far larger than those used in most of the previous researches, including 50,000,000+ records collected from more than 100,000 drives; (2) The ratio of positive and negative samples was highly imbalanced, which was roughly estimated as 1:1000; (3) The signal to noise ratio(SNR) in the SMART data was relatively low, i.e., missing values and measurement errors were widely observed.

Our approach consists of four main steps:

1. Data preprocessing.
2. Feature engineering.
3. LightGBM classification model trained with custom-built focal loss objective function.
4. Failure prediction with a two-step detection rules.

C. He et al. (Eds.): AI Ops 2020, CCIS 1261, pp. 12–17, 2020.
https://doi.org/10.1007/978-981-15-7749-9_2

The LightGBM algorithm was chosen due to the huge size of provided datasets and the algorithm's advantages in terms of computational speed and memory consumption. We could define our custom-build training objective function and evaluation metric in LightGBM.

2 Related Work

Existing SMART-based disk failure prediction algorithms can be mainly classified into the following categories: threshold-based methods, statistical monitoring, binary-classification, and failure time prediction.

The threshold-based algorithms can diagnosis the disk failure with a false alarm rate of 0.1% and fault detection rate 3–10% [1]. The basic idea is that a warning is sent if the normalized value exceeds the threshold. Although the thresholds follow the operator's intuition, they are conventionally tuned to keep a low false alarm rate. In the statistical monitoring, the statistical behaviour of healthy disks is modelled first. Then the deterioration process of disks are estimated by testing whether the data conform to the normal model, examples can be found in [2]. The binary-classification methods such as Bayesian techniques, SVM, linear logistic regression [3–5] use the information from the labels to enhance the accuracy of disk failure prediction based on the SMART-based features. The failure time prediction algorithms make good use of the gradual change in SMART data. They estimate the lead time and examine the health degree [6].

The above works cannot handle the problem of imbalanced dataset very well. Thus, we adopted the LightGBM binary-classifier with focal loss objective function in the competition.

3 The Proposed Approach

3.1 Data Processing

The disk failure date was provided in the train data by the organizers, which can be utilized to label the training samples [7]. As pointed out by [8], disk errors or failure symptoms occur 15.8 days earlier than complete disk failure. That is, the disk samples within a certain amount of days before the disk fault time should be labeled as positive, and the rest samples can be regarded as negative samples. This time interval parameter must be tuned carefully. If the samples far from the fault date are labeled as faulty, or the samples carrying the disk error information are labeled as healthy, additional noise will be introduced artificially.

In our approach, several time intervals were tested based on corresponding prediction accuracy. The days of 30, 15, 10, 7, 5 were evaluated separately, in which 7 was chosen in the end. It should be noted that this evaluation process could be only executed when the whole classification framework was established. This procedure can be illustrated in Table 1.

Table 1. The labeling process.

Disk serial number code	Sampling time	Label
disk_103876	2018-08-22	0
disk_103876	2018-08-22	0
disk_103876	2018-08-23	1
disk_103876	2018-08-24	1
disk_103876	2018-08-25	1
disk_103876	2018-08-26	1
disk_103876	2018-08-27	1
disk_103876	2018-08-28	1
disk_103876	2018-08-29 (fault time of disk)	1

3.2   Feature Engineering

The original SMART data contained 510 attributes, half of them were raw SMART data and the rest were normalized SMART data. An exploratory data analysis showed that only 42 of attributes were non-empty, thus the rest were abandoned.

Then we selected the most relevant SMART attributes from the remaining 42 attributes. This step aimed to discover the attribute set that were most informative predictors of disk failure. The selection was achieved by (1) the change point detection of the attribute data gathered over the time dimension; (2) the application of expert experience in eliminating some irrelevant attributes. This step resulted in a set of attributes with SMART ID of 1, 5, 7, 12, 187, 188, 191, 192, 193, 197, 198, which are listed in Table 2.

Table 2. The SMART attributes used in this paper.

SMART ID	Attribute name
1	Raw read error rate
5	Relocated sector count
7	Seek error rate
12	Power cycle count
187	Reported uncorrectable errors
188	Command timeout
191	G-sense error rate
192	Power-off retract count
193	Load/Unload cycle count
197	Current pending sector count
198	Offline uncorrectable sector count

Instead of adopting the raw feature value, the absolute difference between the current value and its corresponding previous value was utilized. On one hand, many SMART attributes varies along with the runtime, the absolute difference can eliminate this effects to some extend. On the other hand, the absolute difference of SMART attribute time series measures the change of the disk state, which is supposed to characterizes some disk failure symptoms.

Since the sampling time in the train data was one day, one observation on its own was not enough. We aggregated each of the attribute time series to a single value by an exponential moving averaging over a specific time window. Such a feature transformation aimed to improving the numerical stability of the SMART data and assigning more weights to the recent samples.

The above feature engineering processes leaded to 22-dimensional features. We further dropped all the normalized SMART data, resulting in a more compact feature set with only 11-dimensional features. This operation was proved to be effective in improving the final predict results. Here are some explanations. The normalized SMART data are transformed from the raw SMART data, which loses some precision inevitably in the transformation. Besides, some normalized SMART data are highly linearly correlated with the raw SMART data and introduction of the raw data is sufficient enough.

3.3   Objective Function Design

The train data in the scenario of disk failure prediction were highly imbalanced, as only a very small proportion of disks were labeled as faulty. This brought a major challenge to the classification algorithms, as they were typically designed to maximize the overall accuracy. Trained on the imbalanced dataset, the resulting classification algorithms cannot achieve satisfactory performance.

Some practical techniques have been proposed to cope with this issue. First, one can balance the training dataset by downsampling the negative samples or upsampling the positive samples. Common downsampling algorithms can be categorized into generation type and selection type. The former reduce number of samples by generating a new sample set in the targeted class, e.g., representing the original samples with the centroids of K-means method. On the contrary, the latter select samples from the original samples, e.g., randomly selecting s subset for the targeted class. Common upsampling algorithms include naive random over-sampling, Synthetic Minority Oversampling Technique (SMOTE) [9], the Adaptive Synthetic (ADASYN) [10], Generative Adversarial Networks (GAN) [11] and so on. They take advantages of duplication and interpolation to extend the sample number and diversity of the minority class.

In our approach, a new objective function called α-balanced variant of the focal loss was adopted to tackle the problem of imbalanced dataset. This loss function is firstly proposed by [12], which adds a modulating term to the standard cross entropy loss. Such a modification is designed to focus learning on a sparse set of hard examples and prevent the vast amount of easy negatives from overwhelming the classifier during training.

As one can note in Eq. 1, the α-balanced focal loss is modulated with two extra terms: α (or $(1-\alpha)$) and $(1-p)^\gamma$ (or p^γ) when compared to the standard cross entropy loss. By setting α close to 1 (0–1 in practice), the weight of positive samples can be significantly enlarged. Moreover, by setting γ larger than 0 (0.5–5 in practice), the weight of easy samples can be further reduced.

$$\text{FL}(p) = \begin{cases} -\alpha(1-p)^\gamma \log(p)\,, & \text{if } y = 1 \\ -(1-\alpha)\,p^\gamma \log(1-p)\,, & \text{if } y = 0 \end{cases} \tag{1}$$

The application of α-balanced focal loss in the competition was simple and highly effective. Compared with other downsampling and upsampling techniques, we found that replacing the traditional cross entropy loss with focal loss improves F1-score more than 0.05 during many experiments.

3.4   Model

Theoretical analysis and experiments result show that LightGBM can significantly outperform XGBoost and SGB in terms of computational speed and memory consumption. Taking the huge size of train data in this competition into consideration, we decided to use LightGBM as our main algorithm. Readers can refer to [13] and [14] for detailed description. The set of parameters were chosen by grid searching in the parameter space. Our experiments showed that the prediction results are fairly robust to the parameters, e.g., num-leaves, subsample, colsample-bytree, and so on.

3.5   Prediction Logic

A two-step detection logic was proposed to locate the positive samples. In the first step, a probability threshold was estimated from the train set. To be more specific, 0.9997-quantile of predicted probabilities of negatives in the train data was chosen as the threshold value. Thus, in the testing process, the samples will be labeled as positive (i.e., faulty) if their probabilities fall above the threshold. In the second step, only the samples with the TOP N probabilities of each day can be further chosen from the candidates. Note that the number of candidates of each day produced in the first step may be less than the N defined in the second step.

3.6   Conclusion

In our approach, exponentially weighted moving-average of the absolute differences of the 11 raw SMART attribute time series was chosen as the final features. LightGBM model with the focal loss objective function were trained as the underlying classifier. Two-step prediction logic was utilized to obtain the final candidate disks that are predicted to fail within the next 30 days. The final F1-score is 0.4043, while the winning submission score 0.4907. Further research can be focused on collecting different levels of monitoring data except the SMART data, e.g., the logs and monitoring time series from file system, operating system, and the applications that frequently interacting with the disk. If the diversity and heterogeneity of the data are enhanced, more accurate prediction results can be expected.

References

1. Xin, Q., Miller, E.L., et al.: Reliability mechanisms for very large storage systems. In: Proceedings of 20th IEEE/11th NASA Goddard Conference on IEEE Mass Storage Systems and Technologies, 2003, San Diego, CA, USA, pp. 146–156. IEEE Press (2003)
2. Wang, Y., Ma, E.W., et al.: A two-step parametric method for failure prediction in hard disk drives. IEEE Trans. Ind. Inform. **10**(1), 419–430 (2014)
3. Hamerly, G., Elkan, C.: Bayesian approaches to failure prediction for disk drives. In: 18th International Conference on Machine Learning, San Francisco, CA, USA, pp. 202–209. ACM (2001)
4. Murray, J.F., Hughes, G.F., et al.: Machine learning methods for predicting failures in hard drives: a multiple-instance application. J. Mach. Learn. Res. **16**, 783–816 (2005)
5. Yang, W., Hu, D., et al.: Hard drive failure prediction using big data. In: 2015 IEEE 34th Symposium on Reliable Distributed Systems Workshops, Montreal, QC, Canada, pp. 13–18. IEEE (2015)
6. Kaur, K., Kaur, K.: Failure prediction, lead time estimation and health degree assessment for hard disk drives using voting based decision trees. CMC Comput. Mater. Continua **60**(3), 913–946 (2019)
7. The competition datasets. https://github.com/alibaba-edu/dcbrain/tree/master/diskdata
8. Xu, Y., Sui, K., et al.: Improving service availability of cloud systems by predicting disk error. In: Proceedings of the 2018 USENIX Annual Technical Conference, Boston, MA, USA, pp. 481–494. USENIX Association (2018)
9. Chawla, N.V., Bowyer, K.W., et al.: SMOTE: synthetic minority over-sampling technique. J. Artif. Intell. Res. **16**(1), 321–357 (2002)
10. He, H., Bai, Y., et al.: ADASYN: adaptive synthetic sampling approach for imbalanced learning. In: International Joint Conference on Neural Network, Hong Kong, China, pp. 1322–1328. IEEE Press (2008)
11. Salvaris, M., Dean, D., et al.: Generative adversarial networks. In: Deep Learning with Azure, pp. 187–208. Apress, Berkeley (2018)
12. Lin, T., Goyal, P., et al.: Focal loss for dense object detection. In: International Conference on Computer Vision, Venice, Italy, pp. 2999–3007. IEEE Press (2017)
13. Ke, G., Meng, Q., Finley, et al.: LightGBM: a highly efficient gradient boosting decision tree. In: 31st Conference on Neural Information Processing Systems, Long Beach, CA, USA, pp. 3149–3157, Curran Associates (2017)
14. The LightGBM Python packages. https://lightgbm.readthedocs.io/en/latest/Python-Intro.html

A Voting-Based Robust Model for Disk Failure Prediction

Manjie Li[1]($\boxtimes$)(iD), Jingjie Li[2](iD), and Ji Yuan[1,3](iD)

[1] Shanghai East Low Carbon Technology Industry Co., Ltd., Shanghai 200052, China
limanj@elc.cn
[2] Beijing Megvii Co., Ltd., Beijing 100871, China
lijingjie@megvii.com
[3] ERA Group, Technical University of Munich, 80333 Munich, Germany
yuanji1239@gmail.com

Abstract. The hard drive failure prediction is a vital part of operating and maintainance issues. With the fast growth of the data-driven artificial intelligence algorithms, more and more recent researches focus on its application on the current topic. Its effectiveness and powerfulness can be observed through a large number of data experiments. Nevertheless, the prediction accuracy is still a challenging task for dealing with extreme imbalance samples, particularly in big data cases. Rather than merely applying one well-defined LGB model, this study develops a novel ensemble learning strategy, i.e. a voting-based model, for improving the prediction accuracy and the reliance. The experiment results show a progress in scores by employing this voting-based model in comparison to the single LGB model. Additionally, a new type of feature, namely the day distance to important dates, was proven to be efficient for improving overall accuracy.

Keywords: Voting-based strategy · SMART · LGB model · Hard drive failure prediction

1 Introduction

With the fast development of modern cloud datacenters, the number of the hard disk drives deployed has grown dramatically, alongside with the absolute number of disk failures. Since these failures have unneglectable influences on the cloud service quality, the demands of the disk failure detection is increasing as well. Traditional methods mainly follow the rule-based logic by employing SMART (Self-Monitoring, Analysis and Reporting Technology) logs while recent researches show that artificial intelligence algorithm can be a competitive tool to enhance the prediction accuracy and hence gradually becomes a major solution in reality projects. Thereby, for the aim of improving application of AI algorithms, this study builds up a well-defined LGB model and subsequently attempts to

Supported by Alibaba and Shanghai East Low Carbon Technology.

C. He et al. (Eds.): AI Ops 2020, CCIS 1261, pp. 18–29, 2020.
https://doi.org/10.1007/978-981-15-7749-9_3

develop a voting-based strategy. The experiment data source of this project is from 2020 Alibaba AI Ops Competition on Tianchi Platform, i.e. from [1].

For this specific project, comparing to the SMART data collected from other previous applications, it faces to following challenges during the modeling process:

a. Extensive data;
b. Missing records on a daily basis, probably due to hardware or network issues;
c. Difficult to capture the failure status before the failure occurs;
d. Extreme imbalance samples between the healthy and fault disks;
e. Efficient feature construction.

Multiple recent studies have attempted to address aforementioned problems. To facilitate and simplify computation and modeling, extensive data can be splitted into several segments of time series and the most relevant time segment is then chosen for the modeling. Missing records problems are widely distributed in the projects and a natural possible way is to apply filling techniques, e.g. forward and backward filling, liner and nonlinear interpolation methods, etc. Former studies have found that cubic spline interpolation ensures a "smooth" change and achieves a higher TPR (True positive rate) comparing to the spline filling and other methods [2]. Notice that, not many researches specifically emphasized the solution of an imbalanced dataset. For the extreme imbalance cases, naive upsampling and downsampling are potential source of over-fitting [3]. [3] utilized SMOTE (Synthetic Minority Oversampling Technique) for oversampling, yet the precision and recall of the model were decreased.

Except from straightforwardly utilizing given SMART attributes, new features construction greatly influences the prediction accuracy. As a time series problem, statistical sliding window features can be generated to illustrate the distribution of SMART attributes. [4] applied a gradient-based strategy to measure value transitions before disk failures, and its efficiency is validated through the data experiments. A feature combination idea was also brought up in [4], i.e. to take different fault types into account. Nevertheless, it was not clearly presented in [4] how original SMART attributes were combined with new features. Counting the number of attributes that are above zero is another potential approach to combined features [5]. Data from Backblaze shows that when there are more attributes that are above zero (in a certain group), it is more likely to indicate a disk failure.

In general, the mainly contributions of this paper can be concluded as follows: 1) strongly correlated features are extracted based on the data analysis and experiments, i.e. distance to important date, disk usage life, time series slope features and division features; 2) the correlation analysis, which relies on Pearson and Spearman correlation coefficient, is employed for the feature selection, and latter is proved to be more effective in this specific case; 3) a voting-based strategy is developed to ensemble several LGB models with different hyperparameters to improve the accuracy of the disk failure prediction.

2   Feature Engineering

As the basis of a robust prediction model, feature engineering has its irreplaceable place throughout every machine learning process. In this section, various feature processing methods will be illustrated, and their impact on predicting the result will be further discussed in Sect. 4.

2.1   Data Analysis

Dataset provided by 2020 Alibaba AI Ops Competition includes daily SMART logs ranging from July 2017 to July 2018, where all disks belong to a single manufacturer "A" but with different models (model 1 and model 2). The goal of the competition was to predict failure time of disks, and results are evaluated by F1 scores in next 30 days. Thus, for the aim of simplification, the disk would fail or not in next 30 days, denoted as 0 and 1, respectively. Directly afterwards, data exploration was conducted to gain an overall impression on the data.

First of all, Fig. 1 demonstrates failed samples only occupies round 0.08% within the entire train set. It indicates the fact that the train data fed into the model are extremely imbalanced. Undersampling approach, i.e. bagging, was tested on Alibaba's dataset with some modification. Different from other imbalanced problems such as financial fraud where samples are mostly independent from each other, predicting disk failures is more difficult because each model contains continuous data points. To prevent information leak by simply doing bagging on SMART daily logs, bagging on disk serial numbers was applied.

Second, not all original features given are useful for this task. 510 original features were provided by the organizer, i.e. from smart_1 to smart_255. Each attribute possesses both a raw and a normalized value, where the former is measured by the sensors, and relied on the former, the latter is normalized by the manufacturer. For the case with a large number of attributes, attributes selection is full of challenges and arts. Consider the usefulness and completeness of the SMART attributes, we remove features that contains only Nan values and that does not change for all training and testing data samples. Therefore, the dimension of the original attributes is truncated from 510 to 45.

Third, some SMART attribute pairs are highly correlated. As mentioned, normalized attributes are generated by the corresponding raw attributes. It can be inferred that, there could be a strong linear correlation between each pair. Furthermore, some SMART attributes share similar physical meanings and could be non-linearly correlated. This influences might further reach to model training, sometimes negative to some extend, and could lead to potential over-fitting problem. Since they could provide duplicate information. From this perspective, the original feature collection should be restrained by removing strong-correlated features. During the testing, both Pearson and Spearman methods are applied to evaluate the correlation between SMART attributes and labels.

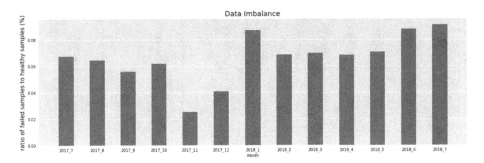

Fig. 1. Samples imbalance situation of the dataset

Fourth, the raw SMART attributes are skewed and needed to be properly transformed. Most machine learning methods are based on the gradient descent algorithm and as known that their performances can be significantly influenced by the given data distribution. Previous data experiments show a distribution, which is close to the Gaussian distribution, usually can provide high stability and good accuracy in comparison to non-Gaussian distributions. To this end, the log-normal transformation was employed for these raw attributes:

$$f' = log(f + 1) \tag{1}$$

in which f' is the feature after log-normal transformation; f is the original feature.

2.2 Feature Generation and Selection

Distance to Important Dates. During the data exploration phase, the distribution of the failed disks was pictured to investigate the trend of disk failures. It was found that more disks would fail on days when important activities were closeby. Understanding the application of the hard disks can shed some light on the feature generation. By summarizing the number of failed disks on certain days (shown in Fig. 2), following conclusions could be drawn: (a) disks were more likely to fail just 1 or 2 days before the starts of the next month (or the ends of the current month); (b) disks were more likely to fail at a few days before important holidays; (c) some peaks can be found after certain holidays. The highest peak observed occurred on Jan 23, 2018, which was about two weeks before the Chinese Lunar New Year. These trends brought a thought that a great portion of the disks were utilized for railway ticket reservation or accounting-related application.

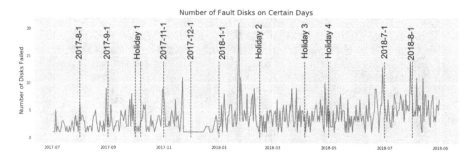

Fig. 2. Trend of failed disks

In Fig. 2, Holiday 1–4 refers to Chinese National Day, Chinese Lunar New Year's Day, QingMing, and Labor's day separately. Among those holidays, the National Day is the longest holiday (7 days). The trend around the National Day shows that peaks appeared before and after the holiday, while maintaining a low level in between. This indicates the fact that whether the day is holiday or not does not matter in this task, but the location relative to the holiday matters. Here a set of new features were proposed: days to next important date, and days to last important date. Important dates were defined based on Chinese holidays and the start of the month in this study. It is worth noting here that researchers should gain some understanding of what disks are used for, since different industries and countries have their own important dates. For instance, it is reasonable for a ticket booking system to observe increased disk failures before holidays but this is not reasonable for a manufacturer quality system due to their different business patterns.

Disk Usage Life. With the disk's usage life increases, the probability of the disk failure becomes higher. Common life span of a hard disk could be in a range of 3–5 years. Although the accumulated training data of Alibaba's disks only last around 1 year and haven't reached normal life end, the disk usage life could still be likely to provide some useful information in failure prediction.

Combined SMART Feature. As discussed in the Introduction section, multiple SMART attributes reaching above zero might indicate a potential disk failure. To validate whether the same trend exists in Alibaba's dataset, the whole data set, i.e. dating from June 2018, was used for exploration. In this study, following group of SMART attributes were selected based on their physical importance: smart_5raw, smart_187raw, smart_188raw, smart_189raw, smart_197raw, and smart_198raw. The result is displayed below in Fig. 3, where x-axis refers to the number of attributes that are above zero. When none of the selected six attributes reaches above zero, 99.95% of those samples are healthy. As the number of larger-than-zero attributes grows, the possibility of failure also rises. Figure 3 indicates a similar trend described in [5], thus can be a potential strong feature in this problem.

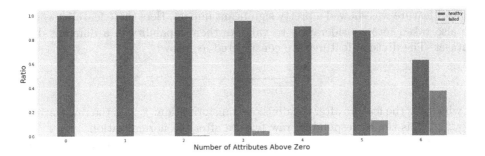

Fig. 3. Trend of sample distribution using combined SMART feature

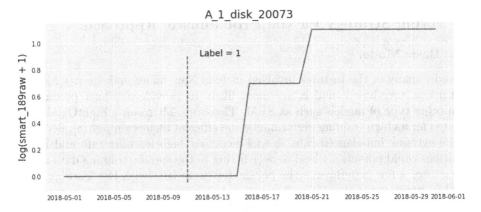

Fig. 4. Smart_189raw change before disk failed

Time Series Slope Features. According to the physical meaning of each SMART attribute [6], many attributes are not real-time but accumulated throughout the running time, such as smart_9 (represents for Power-On Hours). For those attributes, the specific value gives less information than the value changes. Besides, even for attributes that indicate real-time situation, the value changes along the time series could also indicate a status change which relates to disk health. Here one example is shown in Fig. 4: disk 20073 of model 1. The fail date for the disk is June 10, 2018, so based on the label strategy described in Sect. 2.1, all samples after May 11, 2018 (shown as the red dashed line) is labeled as 1. The y axis is the raw smart_189 after the log-normal transformation (see Eq.(1)). The two steps up after May 11, 2018 could be a signal that the disk was close to its life end.

Here in this study, the labeling period was kept as 30 days because of the competition requirement. However, this selection may not apply for all disks.

Division Features. There was another Huawei competition about disk failure prediction which took place the same time as Alibaba's AI Ops Competition, and the champion solution [7] was published online. It was noticed that a new type of feature was showed a pretty significant impact. Here these features would be also taken into consideration to validate their capability in a different disk dataset. The division features are constructed as below:

$$d = \frac{f_{raw}}{f_{normalized} + 0.1} \tag{2}$$

in which d is the feature after the division transformation; f_{raw} is the raw feature; $f_{normalized}$ is the corresponding raw feature after the normalization.

It was not clear enough about the physical meaning of this type of feature, but it does provide some trending information on how linearly the raw and normalized attributes are.

3   Voting Strategy for the Probabilistic Approach

3.1   Basic Model

Consider many of the features involved include Nan values and are not ideally continuous, tree-based models are more likely to provide a robust prediction than other type of models such as SVM. Therefore, Microsoft's LightGBM was selected for its high training performance and efficient memory usage [8]. Because of the extreme imbalanced rate, it was seen that high learning rate and more iterations could potentially lead to over-fitting to the specific training data used. Thus, after a few experiments the parameters were kept with low learning rate and less iterations.

Several down-sampling approach were tested but none of them showed better result. As a result, LightGBM's embedded setting was used to mitigate the imbalance issue. In LightGBM, the parameter "is_unbalanced" provides an approach to deal the imbalance problem with an adjusted loss function [9]:

$$L(y) = -\frac{I_+}{n_+} log P_+ - \frac{I_-}{n_-} log P_- \tag{3}$$

in which denote the modeled conditional probabilities by $P_+ := P(y = 1|x)$ and $P_- := P(y = 0|x)$, and define indicators $I_+(x_i) = 1$ if $y_i = 1$, and $I_-(x_i) = 1$ if $y_i = 0$, vice versa. n_+ and n_- are the number of postive and negative samples, respectively.

Even though the loss function was adjusted for treating imbalanced problem, it was still a logistic approach, and traditional logistic binary classification chooses 0.5 as the threshold to separate negative and positive samples. Nevertheless, with extreme imbalanced dataset used in this task, sticking to 0.5 would categorize almost all samples as healthy. Therefore, a self-defined threshold was put on the calculated probability to better filtering the most-likely failed samples.

3.2   Voting Strategy Framework

Although single LGB model could provide accurate prediction for a group of disk failures, it can also provide completely different result when the combination of training parameters were slighted changed or a different month of data was used for training. There is a possibility that not all LGB models are equally good at predicting all samples, and thus some measures are needed to combine multiple models.

Traditional model ensemble approaches are mostly depended on model blending or stacking, but when investigating the predicted probability for disk samples, totally different probability distribution can be seen for a same set of validation data. This phenomenon implied that predicted probability for failure could be so different that an outlier result can dominate the final result in simple blending or stacking operation. Therefore, we propose a more robust voting strategy to minimize the influences of calculated probability and reliance on a single threshold.

4   Case Study

In order to quantify the effect of features proposed in Sect. 2 and the voting strategy in Sect. 3, multiple cases were tested on Alibaba's docker platform. The models were tested without knowing the specific test data, providing a fairly close-to-real-world environment. Furthermore, we limited the number of disks submitted to 140–160, so that results can be compared without the effect of submissions.

4.1   Attributes Filtering

In Sect. 2.1, it was found that many SMART attributes are highly correlated, therefore, filtering out those high-correlated attributes might bring potential

Table 1. Pseudocode for the voting-based robust model.

Algorithm: The voting-based strategy via a series of LGB models
Step 1: Build up a well-defined LGB model for the binary classification.
i) set the objective function as "binary";
ii) define the metric function as "binary_logloss".
Step 2: Select a series of hyperparameters for LGB models.
i) choose a series of values for hyperparameters, i.e. "learning_rate", "n_estimators" and "subsample";
ii) define combinations of those parameters;
Step 3: Make probability predictions.
For each model, obtain the disks failure probability, i.e.
$Pr_i[y = 1|x], i = 1, 2, ..., m$
in which m is number of models;
Step 4: Define the appropriate threshold.
i) strategy 1: constant threshold θ, e.g. θ =0.005, 0.006, etc.;
ii) strategy 2: adaptive threshold θ, e.g. $\theta = pencentile(Pr_i[y = 1|x], 10)$.;
Step 5: Output disk failure time via a voting-based strategy.
if $Pr_i[y = 1|x] >= \theta$: count += 1
if count >m/2: failure disk;
else: healthy disk.

benefit to the prediction. Table 2 lists the related tests and the corresponding results (Table 1).

Table 2. Result comparison for attribute selection

No.	Method applied	Online score
1a	Single LGB model with non-Nan features	20.57%
1b	Single LGB model using Pearson correlation to filter attributes	19.65%
1c	Single LGB model using Spearman correlation to filter attributes	22.21%
2	Compared to 1a, add distance to important dates feature	21.79%
3	Compared to 1b, add distance to important dates feature	21.88%
11	Compared to 1b, add all generated features and apply voting strategy	22.48%
13	Compared to 1c, add all generated features and apply voting strategy	21.53%

Before additional features were introduced, high-correlated attributes filtered by Pearson correlation slightly reduced the online scores (see 1a and 1b). However, both 1a and 1b indicated the low prediction accuracy and were needed to be improved. This circumstance did not change until the distance to important dates were defined and added as features (compare 1a and 2, 1b and 3). Notice that, with filtering out high-correlated attributes by Pearson correlation presented a little better result than with original attributes (see 2 and 3). This unexpected phenonmenon reveals that the features addition might be nonlinear. Compare to Pearson correlation, Spearman correlation is more suitable for evaluating the statistical dependence between the rankings of two variables and has wider applications. Hence, Spearman correlation test 1c was carried out and a significant improvement can be seen (compare 1b and 1c). Nevertheless, the benefit of the Spearman correlation filtering is not always improving the prediction accuracy (see 11 and 13), it can boost the model under certain conditions but not always the case.

4.2 Features Addition

In Sect. 2.2, five types of features were proposed. To better understand the effect of each feature, 7 tests were conducted step by step in Fig. 5.

The number of features involved were added one by one from No. 1b to No. 8. From the result, it was seen that following constructed features did show a significant contribution: distance to important dates, division features. The influences caused by time series slope features were not clear enough because it lowered the score when comparing No. 5 and No. 6 but improved the score when comparing No. 7 and No. 8.

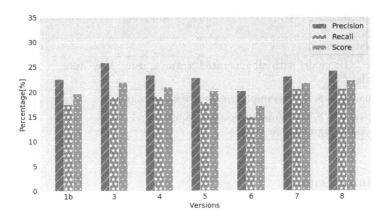

Fig. 5. Result comparison for generated features

No	Method applied	Online score
1b	Single LGB model with only select low-correlated attributes	19.65%
3	Compared to 1b, add distance to important dates feature	21.88%
4	Compared to 3, add hard disk usage life feature	20.94%
5	Compared to 4, add combined SMART feature	20.08%
6	Compared to 5, add time series slope features	17.10%
7	Compared to 5, add corresponding division features	21.68%
8a	Compared to 5, add time series slope features and corresponding division features	22.16%

4.3 Sample Imbalance

As explained in Sect. 2.1, the hard disk dataset is extremely imbalanced. Traditional approach for dealing with imbalance problem includes down-sampling and up-sampling, however, with time-series data as the SMART log data, simple up-sampling and down-sampling approach would result in information leak. Therefore, the approach here was to do bagging or duplication on disk serial numbers instead of simply up-sampling or down-sampling daily records. Result Comparison can be found in Table 3.

Due to the memory limits, a full up-sampling test was not able to be processed. Hence a middle approach was utilized, where down-sampling and up-sampling were combined. It was observed that although these processing method decreased the imbalance extent of the dataset, the model was easily gaining high precision and recall score on the validation set even with low learning rate and less iterations.

Table 3. Result comparison for data up-sampling and down-sampling

No.	Method applied	Online score
8a	Single LGB model with all generated features, using May–June 2018 data	22.16%
8b	Compared to 8, down-sampling negative samples only	17.76%
8c	Compared to 8, up-sampling positive samples and down-sampling negative samples	16.98%

4.4 Training Data

Since the imbalance issue unpreventably brought over-fitting, the selection of the training data becomes essential. Three tests were conducted where only training dataset were varied. The results verifies the hypothesis that training data affects the prediction result to a great extent. This phenomenon raises a concern that there may not exist a set of perfect training data that generates the best result for all testing conditions, meaning it would be hard to know which data should be used for training. However, it was still reasonable to utilize more recent data than data long ago.

4.5 Voting Strategy

As described in Sect. 3, single LGB model does not provide robust results, thus using voting strategy does not only improve the accuracy, but also reduce model's reliance on the probabilistic threshold. In Table 5, it is seen that voting strategy is able to increase the prediction accuracy to some degree, but a bad set of sub-models can also greatly harm the result (Table 4).

Table 4. Result comparison for different training data

No.	Method applied	Online score
8a	Single LGB model with all generated features, using May–June 2018 data	22.16%
9	Compared to 8, use June-July 2018 as training data	14.53%
10	Compared to 8, use July 2018 as training data	17.15%

Table 5. Result comparison for voting strategy

No.	Method applied	Online score
8a	Single LGB model with all generated features	22.16%
11	Compared to 8, use voting strategy instead of a single LGB model: train 7 sub-models with May-June 2018 data	22.48%
12	Compared to 11, add 7 more sub-models trained by June-July 2018 data	17.59%

5   Conclusion

This study investigated various feature construction and filtering methods and proposed a new type of feature, i.e. day distance to important dates. Moreover, a developed voting-based strategy algorithm was applied rather than one well-defined LGB model. During the competition, it was seen that day distance features contributed significant improvement and the voting-based strategy ensured a better result. Noted that, although the tests were designed to gradually add optimized sub-approaches, the benefit of these sub-approaches was not seen to be added linearly. It would be of great worth to investigate the impact of feature combination and interaction. Additionally, this study is worthwhile for further analysis on voting-based approach and can be expanded to include other models, e.g. XGBoost, CatBoost etc.

Acknowledgements. This study and experiment sources are strongly support by the Shanghai East Low Carbon Technology Industry Co., Ltd., and Beijing Megvii Co., Ltd. Thanks to Alibaba, PAKDD for hosting and supporting this competition.

References

1. https://github.com/alibaba-edu/dcbrain/tree/master/diskdata
2. Han, S., et al.: Robust data preprocessing for machine-learning-based disk failure prediction in cloud production environments
3. Aussel, N., Jaulin, S., Gandon, G., Petetin, Y., Fazli, E., et al.: Predictive models of hard drive failures based on operational data. In: ICMLA 2017: 16th IEEE International Conference On Machine Learning And Applications, Cancun, Mexico, December 2017, pp. 619–625. https://doi.org/10.1109/ICMLA.2017.00-92. ffhal-01703140
4. Yang, W., et al.: Hard drive failure prediction using big data. In: 2015 IEEE 34th Symposium on Reliable Distributed Systems Workshops (2015)
5. https://www.backblaze.com/blog/what-smart-stats-indicate-hard-drive-failures/
6. https://en.wikipedia.org/wiki/S.M.A.R.T
7. https://mp.weixin.qq.com/s/LEsJvrB4V3YyOAZP-PGLFA
8. LightGBM Repository. https://github.com/microsoft/LightGBM
9. Burges, C.J.C.: From RankNet to LambdaRank to LambdaMART: an overview. Learning **11**(23–581), 81 (2010)

First Place Solution of PAKDD Cup 2020

Jie Zhang[1](✉), Zeyong Sun[2], and Jiahui Lu[3,4]

[1] Alibaba, Hangzhou, China
zhangjie986862436@gmail.com
[2] Guangdong University of Technology, Guangzhou, China
szy@mail2.gdut.edu.cn
[3] Institute of Computing Technology, Chinese Academy of Sciences, Beijing, China
lujiahui18s@ict.ac.cn
[4] University of Chinese Academy of Sciences, Beijing, China

Abstract. In this paper, we will describe our solution to the PAKDD Cup 2020 Alibaba intelligent operation and maintenance algorithm competition. The biggest challenge of this competition is how to model this problem. In order to maximize the use of data and make model train faster, we turn this problem into a regression problem. By combining GBDT [5] related algorithms like XGBoost [1], LightGBM [2], CatBoost [3,4] and deep feature engineering and utilizing greedy methods for postprocessing the models' predictions, our method ranks first in the final standings with F1-Score 49.0683. The corresponding precision and recall are 62.2047 and 40.5128 respectively.

Keywords: Intelligent operation and maintenance · First place · Regression labelling · Greedy postprocessing

1 Introduction

In large-scale data center, the scale of hard disk usage has reached millions. Frequent disk failures will lead to the decline of the stability and reliability of the server and even the whole IT infrastructure, and ultimately have a negative impact on the business SLA. In the recent decade, industry and academia have carried out a lot of work related to hard disk fault prediction, but there is only a little research on hard disk fault prediction in industrial scale production environment. Large scale production environment has complex business, large data noise and many uncertain factors. Therefore, it has become one of the most important problems that need to be studied and solved in the large-scale data center and the industry in the cloud computing era whether to accurately predict the hard disk failure in advance.

In this competition, we are given a segment of continuously collected (day granularity) hard disk status monitoring data (self monitoring, analysis, and reporting technology; often written as smart) and fault tag data, participants

---

The contribution of three authors to this work is the same.

© Springer Nature Singapore Pte Ltd. 2020
C. He et al. (Eds.): AI Ops 2020, CCIS 1261, pp. 30–39, 2020.
https://doi.org/10.1007/978-981-15-7749-9_4

need to judge whether each hard disk will fail within the next 30 days according to the day granularity. The evaluation indicator of this competition is F1 score, which is usually seen in unbalanced problems.

There are two major challenges for this problem. The first challenge of this problem is how to model. For example,

* The problem of predicting failure can be transformed into a traditional binary classification problem, then we need to predict will the hard disks be damaged or not in the latter 30 days.
* The problem can be transformed into a sorting problem, then we need to predict severity of hard disk damage by learning to rank.
* The problem can be transformed into a regression problem, how many days after will the hard disks be damaged since now.

The second largest challenge of this problem is data noise, imbalance of positive and negative samples, etc.

In next sections, we will explain our methods for solving above problems, experimental results and corresponding feature importance analysis. At method's part, we will show our whole method's details, it contains four steps,

1. Labelling/Modelling strategy, at this part, we will describe the reason why we model this problem as a regression problem and the advantage of this strategy.
2. Feature engineering, since we use GBDT related algorithm as our main models. Feature engineering is core parts for success. At this part, we will describe our feature engineering framework.
3. Models, at this part, we will introduce our selected models, including the reason we choose them and each model's corresponding parameters.
4. Prediction postprocessing strategy, here, we will explain our postprocessing algorithms and its intuition.

After method's part. We will show our method's prediction results and model's feature importance. Finally, we will give a summary of this competition.

2 Method

2.1 Strategy for Labelling

The biggest challenge in this competition is how to model this problem. There are many ways for modelling. The simplest idea is that we can transfer this problem into an binary classification problem. We calculate the difference between the fault time and the current time.

* If the difference between the hard disk failure time and current time is less than or equal to 30 days, we set the label equal to 1.

* If the difference between the hard disk failure time and current time is larger than 30 days. we set the label equal to 0.

The advantage of binary classification labelling strategy is that it is simple and easy to be understood. However, the disadvantage is also obvious: we treat all differences less than 31 days as 1, which makes it harder to distinguish. We know that the difference equals to 30 days may be hard to judge at current time because it is still good and may have no evidence at now, but the difference equals to 1 day may be easy to find some strange indicators. So binary classification labelling strategy makes us lose a lot of important information.

The second strategy is transferring this problem into a multiclass problem. Since the time range is within 31 days. Thus, we can transfer this problem into a 33-class classification problems.

* If the difference between the hard disk failure time and current time is between 0 days and 30 days. We set the label equal to failure time - current time, which is within $[0, 30]$.

* If the difference between the hard disk failure time and current time is larger than 30 days. We set the label equal to 31.

* If the difference between the hard disk failure time and current time is less than 0 days. We set the label equal to -1.

Multiclass classification labelling strategy is much better compared with binary classification. We are able to keep more information compared with binary classification labelling method. The disadvantage of multiclass classification labelling strategy is that it is quite time-consuming if we utilize LightGBM or XGBoost or Catboost for training. Multiclass classification may take several times time compared with binary classification. Besides, it may also lose some information because we set the label of 31 days, 32 days, 33 days ... equal to 31.

The third method is transferring this problem into a regression problem. The method of constructing label is similar to multiclass problems. We use the difference between the fault time and the current time as our label.

* We set the difference between the hard disk failure time and current time as our label.

Regression methods can utilize information better than multiclass classification. We can utilize more information than multiclass when the difference is larger than 30 days. At the same time, we also relieve the problem of time-consuming. Training a regression model is much faster than training a multiclass model in this problem. For this reason, our team decided to use regression labelling strategy as our final strategy.

Note, there exists some differences in our final labelling strategy. During our experiments, we see that if we treat the difference between the hard disk failure time and current time that is larger than 30 days as 31, we can get a little better result. At the same time, we find that if we delete the samples with difference less than -1, we can save some training time while the F1 score does not drop. Concrete labelling strategy can be seen in Table 1. To reduce the influence of noise, we utilize sliding window's methods to construct more training samples, which often gives us better predictions.

Table 1. Regression strategies for labelling

Days to failure	-1	0	1	2	...	30	31	32	33	...
Regression label	-1	0	1	2	...	30	31	31	31	...

After we transferred this problem into a traditional problem, the next step is to make our model do regression better.

In this competition, our team combines deep feature engineering with Light-GBM, XGboost and CatBoost models to solve this regression problem. In the following parts, we will explain concrete details.

2.2   Feature Engineering

Feature engineering is the key to success in many data competitions. Better feature engineering can help our model learn much easier and get better results. At this part, we will explain the details of our feature engineering framework.

Preprocessing. The data set is quite large if we load all data into our machine, which may take more than 100 GB. This may be hard for us to extract other features. Therefore before we do other feature engineering, we need to do some preprocessing first. Here, we do two main operations.

1. Delete feature columns with only null value. Null feature column brings no information, so we can delete them without losing any information.
2. Delete feature columns with only one unique value. Feature column with only one value can not help our model train better. Since the feature in all samples is the same, we can delete them safely.

We can delete 463 features by above operations. To save more memory, we use Memory Reduction Function 1 to do feature types transformation. We save more than 91% memory by doing this, which makes the data size acceptable for us.

Features. At preprocessing part, we make data size acceptable. At this part, we will explain our features. For better explanation, we use V_{now} to represent the feature value of today, and use V_{now-N} to represent the value N days ago.

We split our team's feature framework into five parts, different parts try to capture different information from different aspects.

Algorithm 1. Memory Reduction Function

---

for each feature in all features **do**

    if the type of feature belongs to int **then**

        if min (feature) > min(int8) & max(feature) < max(int8) **then**

            Transfer the type of feature into int8

        else if min (feature) > min(int16) & max(feature) < max(int16) **then**

            Transfer the type of feature into int16

        else if min (feature) > min(int32) & max(feature) < max(int32) **then**

            Transfer the type of feature into int32

        else if min (feature) > min(int64) & max(feature) < max(int64) **then**

            Transfer the type of feature into int64

        end if

    end if

    if the type of feature belongs to float **then**

        if min (feature) > min(float16) & max(feature) < max(float16) **then**

            Transfer the type of feature into float16

        else if min (feature) > min(float32) & max(feature) < max(float32) **then**

            Transfer the type of feature into float32

        end if

    end if

end for

---

1. Meta features: meta features (smart_n_raw and smart_n_normalized) are chosen from original features directly. These features contain disk's original information.

2. Shift features: we do shift operation on original smart_n_raw features, we set the shift days equal to 1, 5, 10, 20 at beginning, which are defined by our experience. Shift features can bring historical information and help our model see further. But to our surprise, we can get better result by only keeping shift day equal to 1 through our experiments.

3. Relative comparison features: we calculate the difference between smart_n_raw and its corresponding shift features, these features can reflect relative change in the last N day.

$$V_{now} - V_{now-N} \qquad (1)$$

4. Absolute comparison features: we calculate the sum between smart_n_raw and its corresponding shift features, these features can relect absolute size. For example, we use A and B to represent today's value and last N day's value, A + B may be quite different though A − B is the same. So, absolute comparison features are complementary to relative comparison features.

$$V_{now} + V_{now-N} \qquad (2)$$

5. Speical Features: we calculate the number of null features of each sample. There are many missing values in our data, which may contain some special informatins. For example, in many loan problems, missing values may tell us that this person does not want to tell us his information. Hence, we use

number of missing values of each sample to represent some special situations here. This helps our model improve around 0.5 to 1 F1-score online.

Concretely, the smart_n_raw features include: 1, 3, 4, 5, 7, 9, 10, 12, 184, 187, 188, 189, 190, 191, 192, 193, 194, 195, 197, 198, 199, 240, 241, 242. and the smart_n_normalized features include: 1, 3, 4, 5, 7, 9, 10, 12, 184, 187, 188, 189, 190, 191, 192, 193, 194, 195, 197, 198, 199, 240, 241, 242. Other features are removed at preprocessing part.

Besides, our team also tried many other features. For example, sliding window's statistical features, we calculate statistical values like mean, median, standard deviation, skewness in the past time windows.

$$Opt(V_{now-M-N}, V_{now-M-N-1}, ..., V_{now-M}) \tag{3}$$

where

* $now - M$: where to start, can be set by yourself;
* N: window size;
* Opt: operations like mean, median, etc.

Ratio features by calculating the ratio between today's value and the value N days ago,

$$\frac{V_{now}}{V_{now-N} + eps} \tag{4}$$

where

* eps: we set 1e−6 here, to prevent the situation when $V_{now-N} = 0$.

Linear coefficient of some feature in the last N days, etc. These features help a lot in many other competitions, especially in time series related problems, but they only bring little gains here, so finally we do not add them into our final models.

2.3   Models

In this competition, we choose LightGBM, XGBoost and CatBoost for training and testing. There are two main reasons we choose these models. The first reason is that there are many null values in our dataset, LightGBM, XGBoost and CatBoost models have nice performance for dealing with these kind of features. The second reason is their outstanding performance in many data competitions. Detailed description of these models can be found in paper [1–4]. The parameter of each model is shown below.

* **XGBoost**: learning_rate = 0.01, n_estimators = 1000, max_depth = 5, subsample = 0.9, colsample_bytree = 0.7, tree_method = 'gpu_hist',
* **LightGBM**: learning_rate = 0.01, n_estimators = 1000, max_depth = 5, subsample = 0.9, colsample_bytree = 0.7, objective = 'mse', random_state = 3,

* **CatBoost**: learning_rate = 0.01, iterations = 1000, max_depth = 5, verbose = 100, early_stopping_rounds = 200, task_type = 'GPU', eval_metric = 'RMSE'.

After many experiments, we find all three models can achieve 44.0+ F1 score online. We take different ensemble strategies, including simple mean operation, weighted mean operation and stacking. To our surprise, we find ensemble with all models can only bring little improvements online. This may be caused by high correlation of prediction results. In the end, our team chooses prediction of XGBoost as our final results.

2.4   Prediction Strategy

Model's prediction can give us a value indicating how many days after will fault happen. There may be many prediction values within 30 days, so we need some strategy for postprocessing and making our final submission. Here, we use greedy strategy to postprocess our model's predictions. The algorithm is shown below 2. Main intuition of our algorithm is that we only consider making prediction today, hence, we need to select the predictions with highest confidence. As we know, if disk's fault day to now is much closer, then the evidence will be much more obvious. Hence, we need to choose samples with highest confidence to submit.

Algorithm 2. Greedy Selection Strategy

---

for model in [1,2] **do**
    Calculate 31 - pred as our confidence value.
    Get count of model at date dt(CountofDt).
    Calculate the rank of confidence value(RankofConf) of corresponding model at date dt.
    If RankofConf $\geq$ p * CountofDt, we set label equal to 1, otherwise, we set label equal to 0, where p is calculated underline.
end for
Submit the date with label equal to 1.

The p value in our algorithm is 0.9983 for model 1 and 0.9985 for model 2. The main disadvantage of our model is that we only choose those samples with highest confidence, so many samples will be ignored. For example, if our model predicts that some disk will fault at 10 days later, but this value may not rank at top, so we will ignore this prediction. But the advantage of greedy selection strategy is also obvious, it is a conservative strategy and in line with our intuitive feelings, we do get high precision and high recall during online testing.

3   Results and Analysis

In the above section, we have explained our method's details, including strategy for labelling, feature engineering, models and greedy selection strategy. In this section, we will show the results of different models we tried and model's feature importance.

3.1 Results

Table 2. Results of different models

Problem transformation	Models	F1 Score	Precision	Recall
Regression	CatBoost	44.3666	52.4138	38.4615
Regression	LightGBM	44.1810	50.9934	38.9744
Regression	XGBoost	49.0683	62.2047	40.5128

As can be seen from Table 2, XGBoost get the highest score in this competition, LightGBM and CatBoost get a little worse results. The main reason is that we tune XGBoost more than the other two models, so there is a little overfitting. We also do binary classification experiments here, but the best F1 score of binary classification is only 41.7943 with precision 57.5221 and recall 32.8205, which again verifies regression labelling strategy's advantage.

3.2 Feature Importance

The gain and split feature importance of XGBoost model are shown in Fig. 1 and Fig. 2 respectively. Here we only show the top 20 most important features. As can be seen from Fig. 1, the top 10 important features include 7 diff features: diff_smart_9raw_1, diff_smart_187raw_1, diff_smart_7raw_1, diff_smart_ 242raw_1, diff_smart_5raw_1, diff_smart_ 193raw_1, diff_smart_ 241raw_1; Two meta features: smart_198raw, smart_199raw; One sum features: sum_smart _198raw_1. This is similar to our understanding, change of recent value is quite an important indicator for predicting disk failure. Absolute features and meta features are complementary to relative features. This is similar to split feature importance, see Fig. 2.

To test the importance of top features, we try to delete top features one by one and see how the score changes. After many experiments, we see that the F1-score may drop from 49 to 43 if we drop the most important feature, here is diff_smart_9raw_1. If we drop any one of the top 10 features, the F1-score all becomes worse. We search for some documents and wiki pages to find these features' physical meanings. We find that smart_9_raw is called power-on time count (POH). Hence, the difference of this feature means the change of power-on time in recent days, which is quite meaningful because we can not say that if POH is large, then the disk will have high probability for failing in the next few days. However, if the power-on time seems unusal in recent one day or two days, then it may be a good indicator that the disk may fail in the next few days.

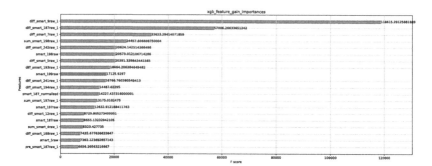

Fig. 1. Feature gain importance of XGBoost Model

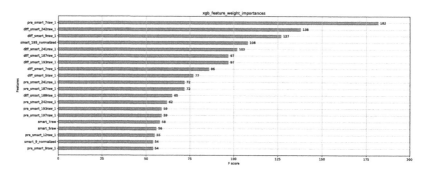

Fig. 2. Feature split importance of XGBoost Model

4   Conclusion

In this paper, we describe our solution to the PAKDD Cup 2020 Alibaba intelligent operation and maintenance algorithm competition. By analyzing three strategies for labelling, our team decided to transfer this problem into a regression problem so that we can maximize the utilization of data information and reduce computation time a lot compared with multiclass classification.

We use feature engineering and XGBoost model as our final training and prediction framework. Our feature engineering framework consists of meta features, relative features, absolute features and special features which helps a lot for model's prediction. Finally, we use greedy selection methods to do postprocessing of our predictions. Our method ranks first in the final standings with F1-Score of 49.0683, with precision 62.2047 and recall 40.5128.

Acknowledgement. Thanks to Tianchi, Alibaba and PAKDD for hosting, creating and supporting this competition.

References

1. Ke, G., et al.: LightGBM: a highly efficient gradient boosting decision tree. In: Advances in Neural Information Processing Systems, pp. 3146–3154 (2017)
2. Chen, T., Guestrin, C.: XGBoost: a scalable tree boosting system. In: Proceedings of the 22Nd ACM SIGKDD International Conference on Knowledge Discovery and Data Mining, pp. 785–794. ACM (2016)
3. Dorogush, A.V., Gulin, A., Gusev, G., Kazeev, N., Prokhorenkova, L.O., Vorobev, A.: Fighting biases with dynamic boosting (2017). arXiv:1706.09516
4. Dorogush, A.V., Ershov, V., Gulin, A.: CatBoost: gradient boosting with categorical features support. In: Workshop on ML Systems at NIPS 2017 (2017)
5. Friedman, J.H.: Greedy function approximation: a gradient boosting machine. Ann. Stat. **29**, 1189–1232 (2001)
6. Disk fault prediction data set. https://github.com/alibaba-edu/dcbrain/tree/master/diskdata

Anomaly Detection of Hard Disk Drives Based on Multi-scale Feature

Xiandong Ran$^{(\boxtimes)}$ and Zhou Su$^{(\boxtimes)}$

School of Cyber Science and Engineering, Xi'an Jiaotong University, Xi'an, China
xiandongran@qq.com, zhousu@xjtu.edu.cn

Abstract. Hard disk drives (HDDs) as a cheap and relatively stable storage tool are widely used by enterprises. However, there is also a risk of fault to the hard disk. Early warning of the HDDs can avoid the data loss caused by the hard disk damage. This paper describes our submission to the PAKDD2020 Alibaba AI Ops Competition, we proposed an anomaly detection method of HDDs based on multi-scale feature. In our method, the original data are classified according to the characteristics of different attributes and proposed a multi-scale feature extraction framework. In order to solve the problem of different data distribution and sample imbalance, the health samples were sampled in time. Finally, we use Lightgbm model to regress and predict the hard disk that will break in the next 30 days. On the real dataset get the 0.5155 precision and 0.2564 recall. Final rank is 24.

Keywords: Anomaly detection · Muti-scale features · Lightgbm

1 Introduction

1.1 Background

With the large-scale increase of data, more and more storage space is needed in the future. Because the HDDs are relatively cheap and stable, many enterprises use the HDDs to store data. However, with the passage of time, the HDDs also have the risk of damage. In case of fault occurred, a large amount of data will be lost. This will lead to irreversible risks. Therefore, it is very necessary to give an early warning to the hard disk which will be damaged soon. The Self-Monitoring Analysis and Reporting Technology (SMART) can monitor the performance of hard disk in real time. Although SMART is widely used, its failure detection rate is low, typically ranging from of 3% to 10% [1].

Supported by Alibaba Clound, PAKDD.

1.2 Related Work

The first time, in 2002–2003, [2,3] use the statistical model based on nonparametric estimation for hard disk fault detection. This method can only be used on hundreds of hard disk samples, which can't be deployed in today's large-scale hard disk anomaly detection scenarios. In [4], the Support Vector Machine (SVM) model is applied to detect the hard disk anomaly, and the prediction accuracy has been further improved. In recent years, more researchers use machine learning and deep learning to solve the problem of hard disk fault warning. Recurrent Neural Network (RNN) is used in [5] for the first time. In [6], they test the various deep learning models in hard disk anomaly detection. Whether it is deep learning or traditional machine learning, these paper do not fully mine the attribute features in SMART. Traditional machine learning uses snapshot features, while deep learning directly uses sequence features. In the actual scene, the extreme imbalance of samples and large number of missing values will also lead to the poor effect of these methods.

1.3 Contribution

With this paper we make the following contributions:

- We analyze the data of SMART in detail, and divide the reasons of the fault hard disk into abnormal value and abnormal trend. Meanwhile, the SMART attributes are filtered by data exploration.
- Not only the snapshot features are used for feature modeling, but also the window features are extracted according to the kind of each SMART attribute. In the experiment, the multi-scale feature improves the prediction effect greatly.
- Under sampling the healthy hard disk samples closest to the prediction date can alleviate the imbalance of samples and greatly reduce the training time. Using regression modeling to make full use of data supervision information.

The remaining of the work is organized as follows: In this paper, firstly, the original data is analyzed in detail and some characteristics of attributes are obtained. In the second section, the specific method of sample selection is proposed. In the third section, we proposed the method of extracting multi-scale features for hard disk anomaly detection. The fourth part and the fifth part respectively introduced the model architecture and experimental details. Finally, the advantages and the future work of this paper are summarized.

2 Data Exploration

2.1 SMART Attribute Exploration

The SMART generates a huge amount of data everyday. In fact, the attributes closely related to the health of the hard disk are limited. Referring to [7,8], 25 attributes are selected as the spare attributes for feature extraction. Furthermore, some attributes with missing value greater than 98% and all value are

Table 1. Some SMART attributes used to extract features.

SMARTID	Meaning	SMARTID	Meaning
4	Start or stop count	5	Reallocated sector count
7	Seek error rate	9	Power-on hours
12	Power cycle count	184	End-to-end error
187	Reported uncorrectable errors	188	Command timeout
189	High fly writes	191	G-sense Error rate
192	Power-off retract count	193	Load cycle count
194	Temperature	195	Hardware ECC
197	Current pending sector count	198	Uncorrectable sector count
199	UltraDMA CRC error	240	Transfer error rate
241	Total LBAs written	242	Total LBAs read

same are eliminated. Finally, we get 20 selected attributes in Table 1. Figure 1 shows the visualization of power on hours, seek error rate and load cycle count (SMART7,9,193). From the visualization of these three attributes, we can see that the value of attributes increases almost linearly with time. If these three attributes are used to extract features, the distribution of training set and test set will be very different. So, it is not advisable to extract the corresponding value as a feature directly for such attribute, which needs to be converted into the relative change amount within a period of time or measured by threshold truncation. In order to find the time distribution of abnormal samples in history, we visualized the number of abnormal samples per month (Fig. 2). According to Fig. 2, it can be found that the number of abnormal samples in each month before 2018 is unstable and there is a large fluctuation. However, the number of abnormal samples close to September 2018 is relatively poor. The number of abnormal samples in 2018 is basically on the rise. The number of abnormal samples in July and August is almost the same. Therefore, training and offline

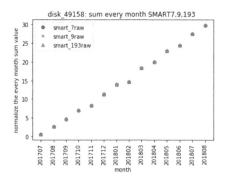

Fig. 1. Monthly sum after normalization

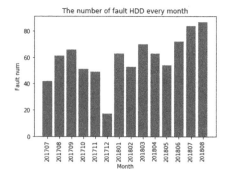

Fig. 2. Total number of bad disks per month

validation should be selected from the data close to September as much as possible. Because the number of abnormal samples is very small, it is easy to get some information for the analysis of abnormal samples. The SMART5 and SMART184 were selected for analysis. Figure 3 shows the change of the attribute value one month before a hard disk is damaged. It can be found that the attribute value began to show an upward trend 30 days before the exception. Therefore, it is necessary to analyze the time series several days before a hard disk failure through multi-scale window. Figure 4 shows the change of SMART184 attribute value 30 days before a hard disk failure. It can be seen that the attribute value of this sample increased abruptly on the day of failure. The occurrence of such outliers needs to be measured by the time snapshot feature.

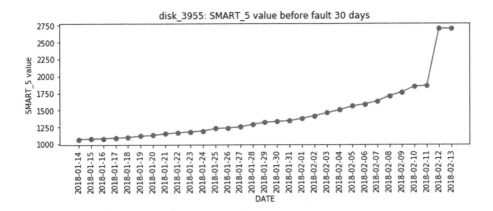

Fig. 3. Change status of the attribute SMART5 before the failure date.

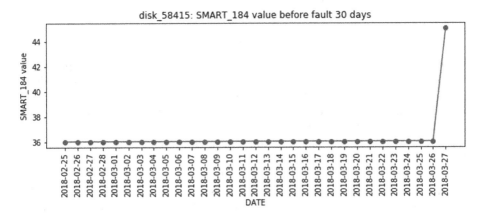

Fig. 4. Change status of the attribute SMART184 before the failure date.

2.2 Build Training Data

The task of this hard disk anomaly detection is to predict whether the hard disk will fail in the next 30 days. The data from July 2017 to August 2018 is used to predict the hard disk data in September 2018. There are two states of hard disk: health or failure. Intuitively, this is a typical binary classification problem. As a binary classification task, it is necessary to provide 0 or 1 label for historical data used for model training. Set the sample label within 30 days from the failure date to 1 (failure) and the rest to 0 (health). However, binary classification will cause a lot of information loss. Each hard disk with a record of fault has the same label for each sample within 30 days or more than 30 days from the date of fault. However, in fact, some SMART attributes have changed on different dates before the hard disk is faulted, and the closer to the damage date, the more obvious the abnormal value may be. Based on this, the problem can be transformed into a regression problem to predict the remaining life of the HDDs. The remaining life of the hard disk can be obtained by calculating the time difference between the current date and the fault date. But, there are a large number of hard disks in the data set which are always in a healthy state, so it is impossible to directly define their remaining life. The piece-wise function can solve this problem, the function is implemented as

$$RUL = \begin{cases} t_w, & t \geq t_w \\ t, & t < t_w \end{cases}$$

where t_w means upper limit of remaining life, t means represents the difference between the date of hard disk failure and the current date. When the t more than t_w or hard drives are always keep health, the label set to t_w. Otherwise, using t itself as label. In this way, all data can be fully utilized.

2.3 Undersampling of Health Samples

The original data size is very large and has a serious sample imbalance problem. Only 0.79% of the total hard disks have failure records. In addition, according to the data visualization in Fig. 1, we have analyzed some characteristics of SMART attributes that grow with time. This also leads to great differences in the distribution of data in different time periods. The attributes of power-on hours, seek error rate and load cycle count grew continuously from 2017 to 2018. The data close to August 2018 is closer to the distribution of real online test sets. Therefore, the solution truncates the data from May 2018 and uses the data after this date as the training set. However, there is still a problem of extremely unbalanced data labels. In our approach, we use the undersampling method for health samples. Directly sample the health samples in July 2018, select 2 million health samples merge to the abnormal samples after May 2018 to get the final training set. In this way, not only the imbalance of samples is alleviated, but also the distribution of samples in the whole month is ensured to be consistent.

3   Feature Engineering

3.1   Feature Classification

Features are divided into two categories: snapshot features and window features. Because different SMART attributes have different characteristics, multi-scale features are needed to measure. According to the first part of data exploration, we can find that the trend of some data will bring the signal of hard disk failure. The other part of the data can reflect the health status of the hard disk from the specific value. Figure 5 shows the overall idea of feature engineering. The time snapshot feature is used for the attribute whose value can reflect the health of the hard disk. The time snapshot feature reflect some specific numerical information. When some attribute values are higher than a certain threshold, the hard disk will be abnormal. But some properties are not sensitive to specific values, and the rising trend may reflect the abnormality of the hard disk. Extracting the statistics feature of a period of time before a certain day can better reflect this trend of change. Of course, some attributes are not only trendy, but also have values that directly affect the health of the hard disk. Then we need to extract window features and time snapshot features respectively.

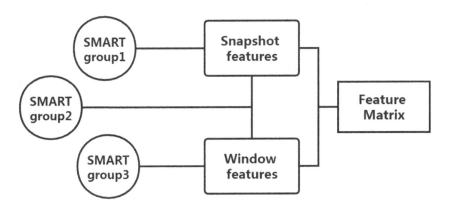

Fig. 5. Feature classification framework.

3.2   Feature Extraction

Snapshot Features. According to the prediction of a certain hard disk one day, its corresponding time snapshot features are divided into two categories. One is the specific value of the SMART attribute corresponding to the target day. The other is the value corresponding to the previous N days. Because some hard disks has no specific value of some SMART attribute at the date of fault time. In this paper, we extract the record values of the previous days. Fig. 6 shows the framework of snapshot feature extraction. Select the days before the date to be predicted to extract the corresponding attribute value. In particular, the missing value is filled with -1.

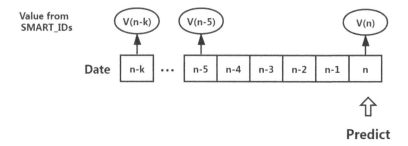

Fig. 6. Time snapshot feature extraction method.

Window Features. The window feature mainly extracts the information within K days before the date to be predicted. The K days before the date to be predicted are regarded as a time series, and the corresponding sequence features are extracted to describe the change of specific attribute values. Many hard disk attribute exceptions can be reflected by time series anomaly. Thus, the problem is transformed into anomaly detection of time series. In the scenario of hard disk anomaly detection, the time series anomaly is mainly manifested as outliers and continuous growth. Some SMART attributes are in a stable state for a long time under normal state. When there is a pulse on the stable sequence, it indicates that there is an abnormality. This anomaly can be reflected by maximum and range. Secondly, the abnormal fluctuation and continuous growth are also typical phenomena reflecting the abnormal of time series. This anomaly can be reflected by variance and sequence mean. Figure 7 shows the window feature extraction framework. In particular, the length of the window can be multiple sizes. This can have a multi-scale effect. The short window can capture the information close to the date to be predicted, while the long window can extract the global information on the long time series.

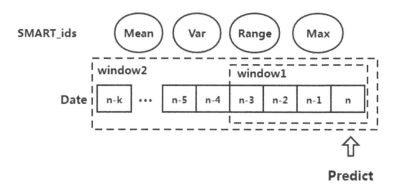

Fig. 7. Window feature extraction method.

4   Model

Lightgbm [9] model is adopted. Lightgbm is the engineering implementation of gradient boosting decision tree (GBDT) algorithm. It adopts the split strategy of leaf wise and histogram difference to optimize. Lightgbm runs much faster than GBDT and is easier to deploy. The data from April 2018 to June 2018 is used as training offline, and the data from August is used for validation. The reason why offline valid data are selected every other month is that when online and September data are predicted, many of the samples in August cannot judge whether they are fault or not. Therefore, the online test data can only be trained with the data from May to July. The number of iteration rounds and offline validation score of the model can be determined by this offline verification method. The method of 5-cross validation is used for online prediction. In this way, the data can be fully trained and relatively stable. Figure 8 shows the overall model architecture. In the cross validation, the model does not converge and needs to rely on the number of training rounds obtained from the time window validation for reference. The average value predicted by five models is used as the final prediction. Select a threshold value for truncation, when the predicted value is less than this threshold output 1, otherwise output 0.

5   Experiment

5.1   Dataset and Evaluation

This paper uses the data provided by Alibaba [10], which comes from the real production. The data includes HDDs data from September 2017 to August 2018. In the test part, the official truncated the test data. The competitors do not know the specific truncation method, which further simulates the uncertainty of the test data on the real production scene. It can better verify the stability of the algorithm. Because HDDs anomaly detection pays more attention to the accuracy of fault samples, the online evaluation function is defined as

$$F1score = 2 \times \frac{Precision \times Recall}{Precision + Recall}$$

the *precision* is defined as $\frac{n_{tpp}}{n_{pp}}$, the n_{pp} means the number of hard disks that are predicted to be damaged in the next 30 days, and n_{tpp} represents the number of hard disks that actually fault in the next 30 days (end September 30) in our predict. The *Recall* is defined as $\frac{n_{tpr}}{n_{pr}}$, then n_{tpr} means the number of hard disks actually fault in the next 30 days (no month limit). n_{pr} represents the number of hard disks fault in the next 30 days in real environment.

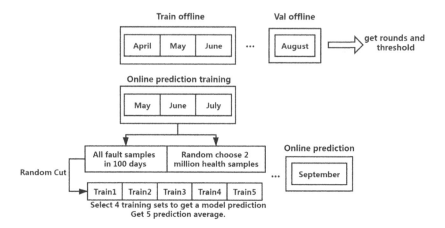

Fig. 8. Modeling and prediction methods.

5.2 Parameter Selection

In order to ensure the speed of extracting window features, we limit the window length to within 30. The window features are extracted by window length of 10, 20 and 30 respectively. Through offline verification, it is found that the verification result is the best when the window length is 20. Refer to [11], we set the piece-wise maximum RUL to be 100. Using the training from March to May and July as the verification, the number of training rounds can be obtained when the model iteration converges is 300. The learning rate of Lightgbm model is 0.03, and the feature fraction is 0.9. The online test data thresholds for the two models are 260 and 285.

5.3 Result

The local model training machine uses 64 g memory and 16 core CPU, and the online prediction environment is 16 g memory CPU. Table 2 shows the online test results under various methods. It can be seen that multi-scale features have obvious advantages over single time snapshot features. Under the same characteristics, the prediction effect of using regression method to model is better than the binary classification method. Through the method of undersampling the health samples, the online prediction performance has been further improved. The winning submission get the 0.5155 precision and 0.2564 recall. Final rank is 24. What's more, it only takes one minute to train the cross validation model offline.

Table 2. Online scores for each method.

Method	Online F1 score
Binary lightgbm with Snapshot features	25.29
Regression lightgbm with Snapshot features ($t_w = 100$)	26.66
Regression lightgbm with multi-scale feature	32.27
Regression with multi-scale feature and undersampling	**34.25**

6 Conclusion

This paper proposes an anomaly detection method based on multi-scale features for hard disk anomaly detection task. Through multi-scale features, the mining of abnormal patterns is more comprehensive. Under sampling of healthy samples can alleviate the imbalance of samples and improve the training speed of the model. The regression method makes full use of the original data and further improves the online score. The Lightgbm model ensures the training speed of the model and is easy to deploy. However, we only extract interval statistical features by sampling time series, we can try to use LSTM to extract time series features in the future work. To solve the problem of sample imbalance, this paper only uses undersampling of healthy samples. In the future, we can try to expand the abnormal samples with GAN, so that the data information can be used more fully.

Acknowledgements. Thanks to Alibaba and PAKDD for hosting, creating and supporting this competition.

References

1. Murray, J.F., Hughes, G.F., Kreutz-Delgado, K.: Machine learning methods for predicting failures in hard drives: a multiple-instance application. J. Mach. Learn. Res. **6**, 783–816 (2005)
2. Murray, J.F., Hughes, G.F., Kreutz-Delgado, K.: Hard drive failure prediction using non-parametric statistical methods. In: Proceedings of International Conference Artificial Neural Network ICANN, Istanbul, Turkey (2003)
3. Hughes, G.F., Murray, J.F., Kreutz-Delgado, K., Elkan, C.: Improved disk-drive failure warnings. IEEE Trans. Rel. **51**(3), 350–357 (2002)
4. Murray, J.F., Hughes, G.F., Kreutz-Delgado, K.: Machine learning methods for predicting failures in hard drives: a multiple-instance application. J. Mach. Learn. Res. **6**(1), 783–816 (2005)
5. Xu, C., Wang, G., Liu, X.G., et al.: Health status assessment and failure prediction for hard drives with recurrent neural networks. IEEE Trans. Comput. **65**(11), 1 (2016)
6. Lima, F.D.S., Pereira, F.L.F., Leite, L.G.M., Gomes, J.P.P., Machado, J.C.: Remaining useful life estimation of hard disk drives based on deep neural networks. In: 2018 International Joint Conference on Neural Networks (IJCNN), Rio de Janeiro, pp. 1–7 (2018)

7. Pereira, F.L.F., Teixeira, D.N., Gomes, J.P.P., Machado, J.C.: Evaluating one-class classifiers for fault detection in hard disk drives. In: 2019 8th Brazilian Conference on Intelligent Systems (BRACIS), Salvador, Brazil, pp. 586–591 (2019)
8. Basak, S., Sengupta, S., Dubey, A., et al.: Mechanisms for integrated feature normalization and remaining useful life estimation using LSTMs applied to hard-disks. In: IEEE international conference on smart computing, pp. 208–216 (2019)
9. Ke, G., Meng, Q., Finley, T.W., et al.: LightGBM: a highly efficient gradient boosting decision tree. In: Neural Information Processing Systems, pp. 3149–3157 (2017)
10. https://github.com/alibaba-edu/dcbrain/tree/master/diskdata
11. Anantharaman, P., Qiao, M., Jadav, D., et al.: Large scale predictive analytics for hard disk remaining useful life estimation. In: International Congress on Big Data, pp. 251–254 (2018)

Disk Failure Prediction: An In-Depth Comparison Between Deep Neural Networks and Tree-Based Models

Xiang Lan[1], Pin Lin Tan[1], and Mengling Feng[1,2]

[1] Saw Swee Hock School of Public Health, National University of Singapore,
Singapore, Singapore
{ephlanx,ephtanpl,ephfm}@nus.edu.sg
[2] Institute of Data Science, National University of Singapore, Singapore, Singapore

Abstract. Disk failures are a constant challenge for data centers, which could lead to data loss or even financial losses. Recently, researchers have proposed predictive models based on SMART attributes. However, most previous studies were often conducted with small-scaled data and usually aimed to predict disk failures just a few hours in advance. This paper aimed at predict if a disk will fail within the next 30 days based on a large scale real world data. The dataset used was from the PAKDD 2020 Alibaba AI Ops Competition, which contains real-world logs from around 170 thousand disks during July 2017 to July 2018. We investigated 3 different kinds of solutions in depth for this task, including time-series prediction, anomaly detection and binary classification. We studied the reasons behind the differences in performance among models, and we proposed a series of data processing methods based on our elaborate feature analysis in order to predict disk failures. Both the offline validation and online testing showed that our observation and proposed methods are promising.

Keywords: Disk failure prediction · Deep learning · LightGBM

1 Introduction

With the rapid growth of information technology (IT), data storage has become increasingly important. This has led to the operation of large-scale data centers that house millions of hard disk drives (HDDs) and solid-state drives (SSDs). Disk failures are a common failure in data centers that could lead to data loss as well as unstable and unreliable servers [8]. They may even affect the entire IT infrastructure, which poses a huge threat to IT companies. Thus, predicting disk failures in advance is crucial for big-data industries.

Self-Monitoring, Analysis, and Reporting Technology (SMART) is a monitoring system in disk drives that reports attributes used to monitor the risk of disk drive failure. A wide range of disk failure prediction methods based on SMART data have been studied [4,6,7,9,10]. The previous studies were often conducted

© Springer Nature Singapore Pte Ltd. 2020
C. He et al. (Eds.): AI Ops 2020, CCIS 1261, pp. 51–63, 2020.
https://doi.org/10.1007/978-981-15-7749-9_6

with small-scaled datasets, and the lead time used to predict disk failures were just a few hours. In this study, we built models that output daily predictions of whether a disk will fail within the next 30 days in the context of large-scale data. The dataset used was from the PAKDD 2020 Alibaba AI Ops Competition and contained real-world data from around 170 thousand disks during 2017-07-31 to 2018-07-31. A more detailed description of the dataset can be found in [1].

To solve the problem of disk failure prediction, we investigated 3 kinds of solutions: Time-series prediction with temporal convolutional networks (TCNs), Anormaly detection with autoencoders and binary classification with gradient boosting decision trees(LightGBM). We comprehensively studied and compared the performance of the models for disk failure prediction task, we also investigate the reasons behind that may lead to the differences in performance among the models.

The main contributions of this paper are: 1) We compared two deep neural network models with gradient boosting decision trees in depth for disk prediction in large-scale data, and analysis the reasons that result in the difference in performance. 2) We proposed a series of data analysis and processing methods based on our elaborate feature analysis for the disk failure prediction task, and the results of the online test showed that our proposed methods are promising.

2   Methods

In this section, we describe our methods including feature analysis, feature engineering, the models we investigated, data preprocessing and the evaluation metrics.

Table 1. Types of variables collected in the raw data.

Name	Type	Description
serial_number	String	Disk serial number code
Manufacturer	String	Disk manufacturer code
Model	String	Disk model code
smart_n_normalized	Integer	Normalized SMART data of SMART ID $= n$
smart_nraw	Integer	Raw SMART data of SMART ID $= n$
dt	String	Sampling time

2.1   Feature Analysis

The raw data included 514 columns containing the 6 types of information shown in Table 1. First, we discarded the SMART attributes that were either all missing or were all the same value in the training set. 52 features remained, including 24 smart_raw attributes and 24 smart_normalized attributes. Note that the methods for calculating smart_normalized attributes are generally vendor-specific and not standardized across different vendors. We visualized the correlation of the remaining features, shown in Fig. 1. The correlation can help in predicting when some attributes are missing. From the correlation matrix we were able to find out some correlated features that could help us to build our model.

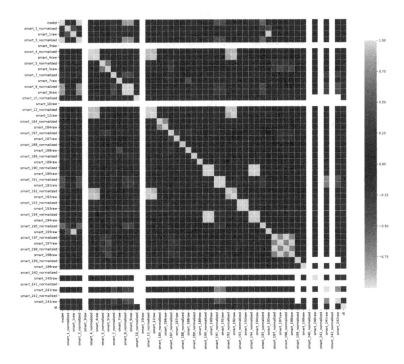

Fig. 1. Correlation Matrix of features shows the correlation coefficients between features, from which we are able to discover features that are correlated to build new features.

The data was labeled as follows: each data point, which corresponds to the status of a disk on a day, was labeled 1 if the disk faulted on the day itself or during the subsequent 30 days. Otherwise the data point would be labeled 0. The data points labeled 1 constitute the positive class, while the data points labeled 0 make up the negative class.

There were two disk model types present in the dataset: disk model 1 and disk model 2. We investigated the difference in the distributions of the features for disk model 1 and disk model 2, as well as for the positive class and negative class. Some examples of the features with different distributions between disk model 1 and disk model 2 and between the positive and negative classes are shown in Fig. 2 and 3 respectively.

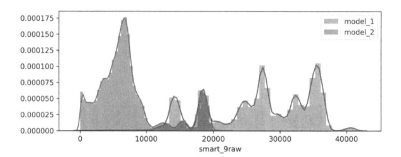

Fig. 2. Distributions of smart_9raw (power-on hours) of disk model 1 and disk model 2. We observed that the distributions of this feature for the different disk models are different, suggesting the heterogeneity between the two disk models.

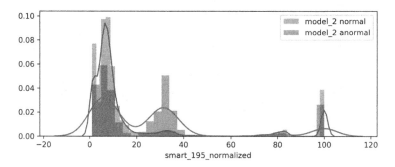

Fig. 3. Distributions of smart_195_normalized (hardware ECC recovered) of disk model 2 disks. It demonstrates that the distributions of the feature are different between normal and abnormal disks, suggesting that we may use this feature to discriminate abnormal disk from normal ones.

The difference in distributions of some SMART attributes between the positive class and the negative class suggests that those features can be used as indicators for a prediction model. However, from our observation, most attribute distributions of disks with the same disk model type are similar between the positive class and the negative class, making it difficult to predict directly from the SMART attributes.

By analysing the feature correlations and testing features on the validation set, we selected 2 sets of features to train our models on.

Feature set 1 consists of the disk model type and the 48 SMART features that were selected by discarding features that are all missing values or all the same value. This set was used as our baseline features.

Feature set 2 is an improved and optimized set of features derived from feature set 1 by first removing the features with a feature importance of 0 according to the LightGBM split gain, this step excluded smart_10_normalized, smart_10raw, smart_12_normalized, smart_188_normalized, smart_191_normalized, smart_191raw, smart_192_normalized, smart_195raw, smart_199_normalized, smart_1_normalized, smart_240_normalized, smart_241_normalized, smart_242_normalized, smart_3raw and smart_4_normalized. Engineered features were then added to the set.

2.2   Feature Engineering

Vertical Feature Engineering: Figure 4 shows the graphs of the smart_5raw (reallocated sectors count) values of 6 different disk before they fault. We can see that the smart_5raw values of the disks tend to increase as they approach their fault dates, but a disk's elevated values may still be lower than another's low values. This suggests that the change in a SMART attribute may be more indicative of a disks probability of fault than the attribute value itself. To make use of this observation, we added features that were the 7-day difference of selected SMART attributes.

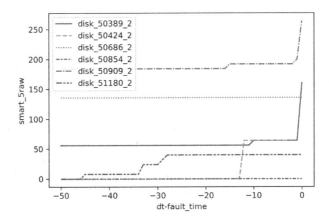

Fig. 4. Graph of smart_5raw (reallocated sectors count) values of 6 different disks in the 50 days leading up to the fault. Although the values vary widely between disks, most of the curves show an upward trend as they approach the fault date, suggesting that the change in value over time may be more important than the attribute value itself.

Horizontal Feature Engineering: One engineered feature we added was the number of days since the disks first log in the dataset. For each disk, we found the date of the earliest log for the disk in the dataset and use the difference between the sampling date and the first date as a feature. It was used as a measure of the usage time of a disk. Meanwhile, from the feature analysis, we found that smart_9raw was a strong feature, and the feature pairs smart_5raw and smart_197raw, smart_7raw and smart_188raw, and smart_3_normalized and smart_191_normalized had correlations that are potential for building new features. For smart_7raw and smart_188raw, smart_5raw and smart_197raw, we calculated the differences between the pairs of features and used the differences as new features.

2.3   Models

Autoencoder: Autoencoding is a data compression and denoising algorithm where the compression and decompression functions are usually implemented by neural networks. The key goal of an autoencoder is to learn a latent representation of the input data from which it can reconstruct the input as accurately as possible. This leads to the autoencoder learning important underlying features.

By using an autoencoder, the task was framed as an anomaly detection task. We trained the autoencoder only on the normal (negative class) data so that the reconstruction loss should be small when the input is normal. Ideally, we would be able to use the loss to classify normal and anormal data points.

We used the same structure for all autoencoders. The structure is simple and composed of three layers: (I) an input layer with size of the number of features, (II) an dense hidden layer of n neurons (n is one third of the input features) with ReLU activation function and L1 norm regularization, and (III) a final dense linear output layer that reconstructs the input. We referred to the autoencoder structure in [3], the autoencoder was trained by using an Adam optimizer to minimize the mean square loss between the input and the reconstruction.

We validated on both normal disks and anomaly disks as we wanted to simultaneously minimize the reconstruction error for normal data points and maximize it for anormal data points. It was important to find a balance. After some experiments, we chose to train the model for 3 epochs using a batch size of 256. This setting prevented the model from over-fitting.

Temporal Convolutional Network: Temporal convolutional networks (TCNs) [2] are a family of convolutional neural networks that handle sequences. TCNs output sequences of the same length as the input sequence, with the nth term of the output sequence calculated from at most the first n terms of the input sequence. TCNs are thus causal; they do not use future data to make the current prediction. The number of previous terms used in each prediction is determined by the size and the dilation factors of the kernels, and the number of layers.

For the experiments, we used the residual blocks described in [2], with the modification of removing the weight normalization. The network used consisted of two residual blocks followed by a convolutional layer. The residual blocks had 64 output channels per layer and kernel size of 4. The dilation factors for the first and second residual blocks were 1 and 2 respectively. The final layer used a kernel size of 1, had 1 output channel and a sigmoid activation function. This results in 19 days of data used for each day's prediction.

The model was trained using the Adam optimization algorithm, using a learning rate of 0.001 and a batch size of 128. Dropout and max-norm constraint was added to all hidden layers. A dropout rate of 0.1 and a max-norm constraint of 2 were used.

LightGBM: LightGBM [5] is a fast, distributed, high performance gradient boosting framework based on decision trees. Unlike other gradient boosting decision tree algorithms, LightGBM uses an enhanced histogram algorithm to divide the continuous eigenvalue to improve the training speed and memory efficiency while maintaining the prediction accuracy. LightGBM uses a leaf-wise generation strategy to reduce the training data which contribute to reduce more losses. An important advantage of LightGBM is its compatibility with large datasets.

For the LightGBM models, the task was framed as a binary classification task. The LightGBM framework we use was the LightGBM 2.3.1 Scikit-learn API. The main parameters of LightGBM are shown in Table 2. In general, the hyper-parameters have a significant influence on prediction accuracy. Hence we used K-folds cross-validation to find the optimal parameters. For training, we used early stopping with the patience set to 50 rounds, using the AUC on the validation set as the metric.

From the cross-validation experiments we determined the num_leaves to 80 or 127 depending on the feature set we used. A higher num_leaves may improve the accuracy but can result in overfitting. The subsample and cosample_bytree hyper-parameters were set to 0.5 to prevent overfitting; learning_rate was set to 0.001, and n_estimators around 100–200 that depended on feature set we used.

Table 2. Main Parameters of the LightGBM model.

Parameters	Interpretation
num_leaves	Maximum tree leaves for weak learners
Subsample	Subsample ratio of the training data
colsample_bytree	Subsample ratio of features that randomly selected when constructing each tree
n_estimators	Number of boosted trees to fit
learning_rate	Learning rate for boosting
max_depth	Maximum tree depth for base learners

2.4 Preprocessing

The disk model and all 48 SMART features that contained values were used to create the inputs to the model. Five features (smart_3raw, smart_10raw, smart_240_normalised, smart_241_normalised and smart_242_normalised) had constant values when not missing. As such, they were converted into features that took a value of 1 when present. The rest of the smart features were normalised such that each had a mean of 0 and a variance of 1.

For the inputs for the autoencoder and TCN models, missing values were filled as follows: the missing values of the five features that had constant values were set to 0 so that they became binary features. The rest of the features' missing values were set to their respective means, which after normalization were all 0. The input into the TCN model needed to be sequences of equal length. Hence, for every disk-date with no data, data was added with all the features except the disk model set to 0. A missing indicator feature for disk-dates was also added.

2.5   Metrics

The precision, recall, and F1-score for the task were defined as follows. For a k-day observation window,

$$\text{Precision} = \frac{n_{tpp}}{n_{pp}}, \tag{1}$$

where n_{pp} is the number of disks that are predicted to fault in the next 30 days during the observation window and n_{tpp} is the number of disks that fault no more than 30 days after the first predicting day in the observation window.

$$\text{Recall} = \frac{n_{tpr}}{n_{pr}}, \tag{2}$$

where n_{pr} is the number of disk faults that occur during the observation window and n_{tpr} is the number of disk faults that occur during the observation window and are successfully predicted no more than 30 days before.

$$F_1\text{-score} = 100 \times 2 \times \frac{\text{Precision} \times \text{Recall}}{\text{Precision} + \text{Recall}}. \tag{3}$$

The k-day observation window for online test was set as 2018-09-01 to 2018-09-30.

Note that the F1-score defined above inherently places a greater importance on the correct prediction of some dates more than others. For example, if a model only predicted that the disk would fault in the next 30 days on the fault-date itself, the model would still obtain a perfect F1-score. On the other hand, if a model only predicted disks to fault in the next 30 days on the day 30 days before the fault, the recall and the F1-score would be 0 on the test set because the observation window is only 30 days long.

3   Results and Discussion

In this section, we analyzed the performance of the different models we used. In particular, we investigated how robust these models are against the distribution shift over time, and we also studied how their performance may vary with different sets of features. We further investigated the reasons of the difference in performance among models, and we use online test to validate these observations.

3.1 Offline Validation Results

For offline validation, we used data from 2018-02-01 to 2018-05-31 (containing 17,748,848 disk logs) for training and data from 2018-07-01 to 2018-07-31 (containing 4,855,732 disk logs) for validation. Note that data from 2018-06-01 to 2018-06-30 was not included in the training set. This is because using labeled data from June 2018 may cause data leakage; both labeling the data in June 2018 and the validation set use the information of the disk faults that occur in July 2018. This replicates the online test setting, where we have to predict for September 2018, for which we have no disk fault information. The validation results are displayed in Table 3.

Table 3. Maximum F1-scores of the models on the validation set

Model	Baseline	Feature Set 2
TCN	17.31	19.08
LightGBM	19.95	24.25
Autoencoder	6.10	9.80

From the validation results we can see that the performance of all models improved after feature engineering, and that the LightGBM was the most feature sensitive. The autoencoder performed much poorer than the TCN and LightGBM models who had comparable performances. Despite the fact that the LightGBM model only used one day of data while the TCN model used 19 days of data, the LightGBM model outperformed the TCN model on both feature sets on the validation data.

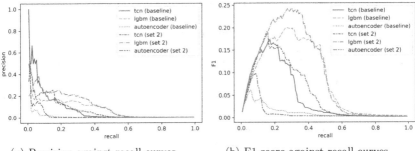

(a) Precision against recall curves.   (b) F1-score against recall curves.

Fig. 5. The precision-recall and F1-recall curves of the different models. The TCN models have the higher precision than the other models at low recall scores, while the LightGBM models have higher precision scores at higher recall scores. Because the models achieve maximal F1-scores at recall scores where the LightGBM models have higher precision, the LightGBM models have higher maximal F1-scores.

To explore the differences in the models, we plotted the precision-recall and the F1-recall curves for the models on the validation set, shown in Fig. 5. The graphs show that the precision of the TCN models is higher than that of the LightGBM models at lower recall scores (Fig. 5(b)) but lower at higher recall scores. Because the high F1-scores occur at the recall values where the TCN models have lower precision scores, the TCN models have lower maximum F1-scores.

Another observation we made was that the thresholds used to achieve the maximum F1-score were very low. The best thresholds for the TCN models were in the range 0.076–0.109, while the thresholds for the LightGBM models were in the range 0.001–0.007. This suggests that the cross-entropy loss function may not be a very suitable loss for optimizing the F1-score. If the cross-entropy loss function was well-suited for optimizing the F1-score, we would expect the optimal thresholds to be nearer to 0.5.

A possible reason for the poor match between cross-entropy loss and the F1-score is that the cross-entropy loss gives equal weight to the loss of each data point. However, as previously mentioned, the F1-score defined above rewards the correct prediction of some days more than others. A possible remedy to this problem would then be to use a weighted cross-entropy loss, where the weight of a data point depended on how many days it was away from the fault date. Data points of a disk on the day of its failure would be given a high weight while day data points far from the fault date would be given lower weights.

Meanwhile, an interesting observation is the comparison between the two neural networks models; the TCN outperformed the autoencoder. One major difference between the two models is that TCN used convolutional layers to extract features, while the autoencoder was composed of dense layers, in which each neuron is fully connected. The dense layer is learning a representation of all features. This process may produce some useful latent features but may also produce many unrelated latent features that confuse the model for prediction. This is reflected in the overall performance and the improvement after adding new features. We also explored different autoencoder structures by adding hidden layers to make it deeper and change the size of hidden layers, and most structures shows an similar performance with the baseline autoencoder. Also, the autoencoder is unable to learn features that occur over time, while the TCN models uses of 19 days of data.

We tried using an customized oversampling algorithm to counter the impact of data imbalance. We oversampled the positive class, with data points closer to the fault date sampled a greater number of times than those further away. This reflected how the F1-score placed a greater importance on predicting dates close to the fault date. However, oversampling did not improve the performance of the models in our experiments, so we did not use it further. More sophisticated oversampling algorithms for this task are promising to be studied in the future.

To investigate other potential reasons, we visualized the latent representation from the hidden layer of the autoencoder in 2-dimensional space using t-

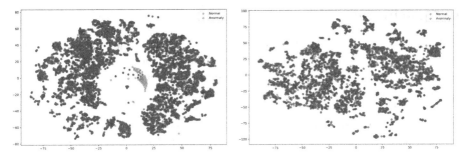

(a) latent representation on training data (b) latent representation on validation data

Fig. 6. Latent representation from autoencoder. (a) shows the latent representation from hidden layers on training data, It's clear that the latent representation in the training set is partly separable with data points overlapping mostly in the left portion of the plot. (b) shows the latent representation from hidden layers on validation data that most of the data points overlap. For observations above, they imply that the autoencoder model only learned to fit the training data.

distributed stochastic neighbor embedding (t-SNE) of the autoencoder, shown in Fig. 6.

3.2   Online Test Results

The online test is to prove our observation and conclusion from the experiments results in validation set, we select our models with feature set 2 for online test, we controlled the features and data processing methods are the same for each model, and the threshold is selected from validation, the date range of training data from 2018-04-01 to 2018-07-31, the organizer did not provide the fault tag for 201809, so we skip 201808 for training, and date range of the online test from 2018-09-01 to 2018-09-30. The results shown in Table 4.

Table 4. Online test results. The LGBM shows a close performance to our validation, supporting our conclusion for these three models from validation and showing our proposed methods are promising. However, TCN shows a significant difference from the performance on the validation set. A reason is that we did not adjust the threshold for the online test. The threshold is from the best observation in validation. This observation implies that deep neural network model are more sensitive to the change of distribution of data, while gradient boosting decision trees are more stable and robust.

Model	Precision	Recall	F-Score
TCN	66.67	1.02	2.02
LGBM	23.79	30.77	26.83
Autoencoder	19.70	6.67	9.96

4 Conclusion

In this paper, we investigated three different solutions for the task of predicting disk failures that occur within the next 30 days under the setting of the PAKDD 2020 Alibaba AI Ops Competition. We comprehensively compared deep neural networks models to gradient boosting decision tree based model and studied the potential reasons behind that lead to the difference in performance. We found that a gradient boosting tree-based model (LightGBM) was more suitable and robuster for this task than the other two deep neural network models. For this task, the deep neural models were more sensitive to the changes in the data distribution and may resulting in poor performance. Meanwhile, we proposed a series of data analysis and processing methods for the disk failure prediction task, and the results of the online test showed that our proposed methods are promising.

Acknowledgements. We would like to thank the organizers for organizing such a meaningful competition, for all the preparations for this competition and their patience in answering the questions raised by the contestants. This research is supported by the National Research Foundation Singapore under its AI Singapore Programme [Award No. AISG-GC-2019-002], the NUHS Joint Grant [WBS R-608-000-199-733] and the NMRC Health Service Research Grant [HSRG-OC17nov004].

References

1. https://github.com/alibaba-edu/dcbrain/tree/master/diskdata
2. Bai, S., Kolter, J.Z., Koltun, V.: An empirical evaluation of generic convolutional and recurrent networks for sequence modeling. arXiv preprint arXiv:1803.01271 (2018)
3. Borghesi, A., Bartolini, A., Lombardi, M., Milano, M., Benini, L.: Anomaly detection using autoencoders in high performance computing systems. In: Proceedings of the AAAI Conference on Artificial Intelligence, vol. 33, pp. 9428–9433 (2019)
4. Kaur, K., Kaur, K.: Failure prediction and health status assessment of storage systems with decision trees. In: Luhach, A.K., Singh, D., Hsiung, P.-A., Hawari, K.B.G., Lingras, P., Singh, P.K. (eds.) ICAICR 2018. CCIS, vol. 955, pp. 366–376. Springer, Singapore (2019). https://doi.org/10.1007/978-981-13-3140-4_33
5. Ke, G., et al.: LightGBM: a highly efficient gradient boosting decision tree. In: Advances in Neural Information Processing Systems, pp. 3146–3154 (2017)
6. Li, J., et al.: Hard drive failure prediction using classification and regression trees. In: 2014 44th Annual IEEE/IFIP International Conference on Dependable Systems and Networks, pp. 383–394. IEEE (2014)
7. Pang, S., Jia, Y., Stones, R., Wang, G., Liu, X.: A combined bayesian network method for predicting drive failure times from smart attributes. In: 2016 International Joint Conference on Neural Networks (IJCNN). pp. 4850–4856. IEEE (2016)
8. Vishwanath, K.V., Nagappan, N.: Characterizing cloud computing hardware reliability. In: Proceedings of the 1st ACM Symposium on Cloud Computing, pp. 193–204 (2010)
9. Wang, Y., Miao, Q., Pecht, M.: Health monitoring of hard disk drive based on mahalanobis distance. In: 2011 Prognostics and System Health Managment Confernece, pp. 1–8. IEEE (2011)

10. Xiao, J., Xiong, Z., Wu, S., Yi, Y., Jin, H., Hu, K.: Disk failure prediction in data centers via online learning. In: Proceedings of the 47th International Conference on Parallel Processing, pp. 1–10 (2018)

PAKDD2020 Alibaba AI Ops Competition: Large-Scale Disk Failure Prediction

Run-Qing Chen[✉]

Department of Computer Science, Xiamen University, Xiamen, China
`chenrq1010026261@stu.xmu.edu.cn`

Abstract. The failure prediction for storage systems plays more and more important role given the explosive growth of data in various data centers in recent years. In this paper, the existing technologies have been employed in the prediction have been reviewed. In particular, the techniques such as imbalance data alleviation and temporal feature construction, which are also adopted in our solution, are reviewed in more detail. Our solution to the prediction problem which is mainly built upon Light-GBM is then presented. The solution ranks 38 with F_1-score of 34.28% on PAKDD2020 Alibaba AI Ops Competition.

1 Introduction

With the explosive growth of data, the security of the storage center responsible for data storage has gradually received more and more attention in recent years. Once the storage system fails suddenly, a lot of resources will be consumed in system maintenance, data migration and data recovery. Therefore, the failure prediction of disks is a critical and fundamental task. Most disk failures are a slow deterioration such as mechanical wear due to their physical structure. These failures are believed predictable according to some indications, which may include the number of damaged disk sectors, heat output and so on. Predicting these failures in advance allows operation engineers sufficient time for maintenance and replacement, and greatly reduces the resources for data recovery.

In the literature, the related work explores each procedure of the task, such as data collection [1,13,21], feature selection [2,9,11,12,18,21,22], temporal feature construction [2,20–24], feature preprocessing [11,18,19,22,24], data imbalance alleviation [2,16,17,24], model for classification/regression [2,6,7,9,12,13,15,17, 18,22–24], model update [9,19,22] and evaluation metrics [10,11]. Although the above explorations seem to show good performance on some datasets, it may be not applicable in large-scale datasets. This is because there are large data noise and many uncertain factors in large-scale datasets. In this paper, analysis and discussion about some explorations are made, combined with our practice in the PAKDD Alibaba AI Ops Competition [4].

© Springer Nature Singapore Pte Ltd. 2020
C. He et al. (Eds.): AI Ops 2020, CCIS 1261, pp. 64–73, 2020.
https://doi.org/10.1007/978-981-15-7749-9_7

2 Related Work

SMART which stands for Self-Monitoring, Analysis, and Reporting Technology is a monitoring system installed with the computer disk drives. There are *255* monitoring attributes/statuses are provided by SMART. These statuses including temperature, power-on hours, write error rate, etc. Each status is given both as a raw value and a normalized value. SMART is used to detect and report various statuses, based on which failure prediction is made. However, simple thresholding on these statuses only results in very poor prediction precision. Usually, only *3–10%* failure detection rate (FDR) is achieved when we control the low false alarm rate on the order of *0.1%* [14].

Feature Selection. With the raw data from SMART, it is important to select features for the prediction model as not all of the attributes are relevant to disk failure. Pruning the redundant features also leads to the better generalization of the prediction model. Failure modes, mechanisms, and effects analysis (FMMEA) are introduced to identify which attributes are the most relevant to each type of failure [18]. Moreover, three non-parametric statistical methods (reverse arrangement test, rank-sum test and z-scores) are adopted in [9, 11, 12]. Compared with statistical methods, the relevant attributes are selected directly by the detection of permanent changepoints [2]. To select attributes with long-term predictability, online learning is adopted in [21] instead of cross-validation.

Temporal Feature Construction. In order to incorporate the other information such as the trend of a period into the prediction model, efforts have been also taken in the construction of temporal features. For instance, Hidden Markov models (HMM) and Hidden semi-Markov models (HSMM) are adopted to capture status trends and temporal dependencies [23]. Furthermore, some statistical features such as difference and variance are calculated for failure prediction in [21, 22, 24]. In order to capture the long-term dependencies in the sequence, recurrent neural network (RNN) [20] and exponentially weighted moving average model (EWM) [2] are used in disk failure prediction.

Data Imbalance Alleviation. In the real-world situation, the failed disks in rare occurrences result in data imbalance, which greatly damages the performance of the model. To address this issue, different sample rates are used for healthy disks and failed disks [17, 24]. To choose the more representative subset from healthy disks, *k*-means is adopted to cluster healthy disks and select the data points closest to the respective cluster centroid [2]. Moreover, to get a more balanced subset of category distribution, one-sided selection is used to remove the redundant samples and the noise samples [16].

Model. In the literature, disk failure prediction is addressed as binary classification between healthy and failed disks or regression of a gradual deterioration from healthy to failed. The advantage of regression-based approach is that the health degree of the disk instead of the status of healthy or failed is given. The basic models are adopted in the early work, such as Mahalanobis distance [18], support vector machine (SVM) [23], logistic regression [22], decision tree [9], and neural network [24]. Due to the superior performance of ensemble learning, combined Bayesian network (CBN) is adopted to combine four base classifiers to get better prediction in [15]. In addition, tree-based ensemble learning such as gradient boosted regression tree (GBRT) [10,12], random forest (RF) [7,13,17], and regularized greedy forest (RGF) [2] is adopted, since it has excellent interpretability with good performance.

3   Our Approach

For each disk, the participants are provided the manufacturer code, the model code, the serial number code, and a period of time of daily SMART data and the fault label. The disks with the same serial number code and the different model codes are different. The disks are therefore assigned with identity code as "the model code+the serial number code". The task is to predict daily whether one disk will fail within the next *30* days. In our solution, it is modeled as a binary classification. In the training, the data of *30* days before the disk failure happens are treated as positive, while the rest are labeled as negative data.

According to the related work [2], SMART data from different manufacturers and different models demonstrate different distributions. The prediction performance will be degraded if it is based on simple information integration over them due to the big distribution difference. In this competition, all the manufacturer codes are '*1*', which indicates that all the disks are from one manufacturer. The model codes are '*1*' or '*2*', which indicates that there are two models in the dataset. The disks of these two models are combined in the training, since the distributions of the two models are similar according to observation.

The supplied data from the competition committee are noisy. The three typical cases that the data are polluted are listed as follows.

- There are a lot of missing attributes, such as soft read error rate, erase fail count.
- Some attributes have only one or two values, whose information for prediction is small.
- The normalized value of some attributes is the same as the raw value, which is redundant for prediction.

To address these issues, the attributes with more than *60*% of missing data and the attributes with few unique values are removed. Moreover, the normalized values that are equal to the raw values are removed. The data attributes after the removal of noisy data are summarized in Table 1.

In order to incorporate the temporal features as much as possible, feature selection is adopted only after all the temporal features are produced.

Table 1. The attributes supplied by SMART

ID#	Attribute name	Raw	Norm
1	Read error rate	✓	✓
3	Spin-up time		✓
4	Start/Stop count	✓	✓
5	Reallocated sectors count	✓	✓
7	Seek error rate	✓	✓
9	Power-on hours	✓	✓
10	Spin retry count		✓
12	Power cycle count	✓	✓
184	End-to-end error	✓	✓
187	Reported uncorrectable errors	✓	✓
188	Command timeout	✓	✓
189	High fly writes	✓	✓
190	Temperature difference	✓	✓
191	G-sense error rate	✓	✓
192	Power-off retract count	✓	✓
193	Load cycle count	✓	✓
194	Temperature	✓	
195	Hardware ECC recovered	✓	✓
197	Current pending sector count	✓	✓
198	Uncorrectable sector count	✓	✓
199	UltraDMA CRC error count	✓	✓

According to [21], given a time window w, difference and variance are calculated with Eq. 1 and Eq. 2.

$$\text{Diff}(x, t, w) = x(t) - x(t - w), \qquad (1)$$

$$\text{Sigma}(x, t, w) = E[(X - \mu)^2], \qquad (2)$$

where $X = (x_{t-w}, x_{t-w-1}, \ldots, x_t)$ and $\mu = \frac{\sum(X)}{w}$. Different sizes of the time window are adopted for calculating difference and variance in our approach. Due to the large dataset and the few failed disks, other temporal features are not constructed. On the one hand, the computational complexity of RNN, HMM and least squares is high, resulting in the long construction time. On the other hand, EWM over-compacts the temporal representations, resulting in a smaller amount of trainset.

Since tree-based LightGBM [8] is adopted for binary classification, the normalization is unnecessary. It is also unnecessary to perform discretization due to the histogram optimization of LightGBM.

In order to alleviate the data imbalance, the following schemes are tested.

- To undersample the healthy disks, k-means is adopted to cluster the healthy disks and the several nearest disks from each cluster centroid are chosen as the subset.
- To synthesize the failed disks, Synthetic Minority Oversampling Technique (SMOTE) is adopted to select a sample b from the nearest neighbors of a sample a, and then randomly choose a point on the line between a and b as the synthesized sample.

Unfortunately, the above schemes are found not helpful. The performance from the model with the above schemes is presented and analyzed in Sect. 4.

In this scenario, the model is trained with the samples of the current time period, and then tested in the samples of the next time period. According to [5], there are some features available in the current time period, but not helpful in the next time period due to the concept drift of time series. To address this issue, a feature pruning based on [21] is tested, which simulates this scenario in the trainset. In order to save time and memory, the feature is selected according to their importance of LightGBM. The trainset is split by time into two parts for training and validating. If the performance in the validation set is better after the removal of the features with the least importance, the features are removed until the performance is no longer better. And the result is in Sect. 4.

After feature pruning, LightGBM [8] is trained as a binary classifier. Compared with XGBoost [3], LightGBM is fast in training due to the histogram optimization and leaf-wise technology. The hyper-parameters are determined by performing a simple grid search in the parameter space.

4   Experiments

In this section, different configurations of our solution are studied on the dataset released by the organization committee of the competition. The trainset is collected from July 2017 to August 2018, and the brief information is summarized in Table 2. To prevent the model from degrading in the testset, the data in June 2018, July 2018, and August 2018 are chosen as our trainset. And about 200 prediction results with the highest probability of failure after deduplication are submitted. In the evaluation, only the earliest predicted date of failure for each disk is considered if there are multiple prediction results for a single disk. According to the purpose of failure prediction that predicting whether each disk will fail or not within next 30 days, the precision, recall and F_1-score are redefined as Eq. 3, Eq. 4 and Eq. 5 for the evaluation.

$$\text{precision} = \frac{n_{tpp}}{n_{pp}}, \tag{3}$$

$$\text{recall} = \frac{n_{tpr}}{n_{pr}}, \tag{4}$$

$$F_1\text{-score} = 2 \times \frac{\text{precision} \times \text{recall}}{\text{precision} + \text{recall}}, \tag{5}$$

where n_{tpp} is the number of all the disks those truly fail among 30 days after the first predicting day in the observation window, and n_{pp} is the number of disks that are predicted to be faulty in the following 30 days. n_{tpr} is the number of truly faulty disks that are successfully predicted no more than 30 days in advance, and n_{pr} is the number of all the disk failures occurring in the observation window. The observation window in the semi-final is from 20180901 to 20180930.

Table 2. Summary over the trainset

Period	# Samples	# Disks	# Failed Disks
201707	1,610,007	102,732	107/0.10%
201708	3,140,857	124,624	155/0.12%
201709	3,146,965	119,106	129/0.11%
201710	3,536,616	134,252	142/0.11%
201711	3,639,220	140,757	99/0.07%
201712	4,105,046	143,397	121/0.08%
201801	4,428,248	148,149	230/0.16%
201802	4,083,439	149,388	208/0.14%
201803	4,566,413	151,178	218/0.14%
201804	4,463,347	152,204	204/0.13%
201805	4,647,989	153,960	229/0.15%
201806	4,647,434	163,106	279/0.17%
201807	4,855,732	167,184	294/0.18%
201808	5,161,737	176,369	161/0.09%

4.1 Results

LightGBM without any modification is treated as the comparison baseline in the evaluation. The runs that are additionally integrated with imbalance data alleviation and temporal features are evaluated. By this way, we try to investigate the contribution of each component to the prediction accuracy.

Table 3. Performance of difference feature (Diff) with different window sizes

Approach	F_1-score	Precision	Recall
Without diff	24.82	24.52	25.13
With diff3	**34.28**	**34.72**	**33.85**
With diff7	32.95	34.81	31.28
With diff15	29.59	31.76	27.69

Temporal Feature Construction. The temporal features are derived from the difference and variance of a time window. The performance from difference features produced with different window sizes is presented in Tab. 3. As shown in the table, the performance of the difference features with a window size of 3 is better than other window sizes and the baseline. Specifically, more than 9% improvement is observed on the performance with a window size of 3, compared with the baseline. Therefore, the difference feature with a window size of 3 is used in the following experiments. In addition, the performance of variance feature with different window sizes is summarized in Tab. 4. As shown in the table, the variance feature brings no improvement than that without variance. As a result, the variance feature is not used in the following experiments.

Table 4. Performance of variance feature (Sigma) with different window sizes

Approach	F_1-score	Precision	Recall
Without sigma	**34.28**	**34.72**	**33.85**
With sigma3	34.09	33.82	34.36
With sigma7	34.17	34.50	33.84
With sigma15	31.54	31.79	31.28

Data Imbalance Alleviation. To alleviate the data imbalance, k-means is adopted to undersample the healthy disks. k is set to 112 and 100 disks closest to each cluster centroid are chosen as the subset. In addition, another attempt, SMOTE is adopted to synthesize the failed disks until the number of healthy disks and failed disks are equal. The performance of these two attempts is summarized in Table 5. As shown in the table, there is a large performance gap between the baseline and these two attempts. The performance of k-means is not improved but reduced. This is because the information of the unselected samples is totally removed, although the selected subset contains the most representative samples. Moreover, there is no performance improvement in the attempt with SMOTE. And the probability threshold of failed disks is found to be increased. The reason may be that the information of the synthesized disks overlaps with that of the original failed disks, which even degrades the model in the training. Therefore, these two attempts are not used in the following experiments.

Table 5. Performance from imbalance data alleviation

Approach	F_1-score	Precision	Recall
Without allevation	**34.28**	**34.72**	**33.85**
With k-means	1.45	1.37	1.54
With SMOTE	18.30	17.26	19.49

Feature Selection. To prune the feature without long-term effectiveness, an approach bases on [21] is attempted to prune the unimportant features first. The performance of feature selection is summarized in Table 6. As shown in the table, there is also no improvement in this attempt, although it demonstrates better performance in the offline validation set. This may be because the distribution difference between the offline validation set and the online testset is large. As a result, the attempt is not used in the final approach.

Table 6. Performance of feature selection

Approach	F_1-score	Precision	Recall
Without selection	34.28	34.72	33.85
With selection	33.17	33.00	33.33

Comparison with Other Solutions. The performance of our solutions and top solutions of the leaderboard on the dataset from the competition is summarized in Table 7. As shown in the table, there is a large performance gap between the solution of rank *1* and other solutions. And the performance of our solution ranking *38* is not far from that of the top solutions of the leaderboard. As far as we know, the solutions of rank *1* and rank *2* are based on regression, which will be our attempts in the future.

Table 7. The comparison with other solutions

Approach	F_1-score	Precision	Recall
Rank *1*	49.07	62.20	40.51
Rank *2*	42.51	53.24	35.38
Rank *3*	40.47	52.76	32.82
Rank *4*	39.98	52.42	32.31
Rank *5*	39.71	43.68	36.41
Our solution	34.28	34.72	33.85

5   Conclusion

In this paper, we have experimented and analyzed some explorations such as temporal feature construction and data imbalance alleviation, in combination with our attempts in PAKDD2020 Alibaba AI Ops Competition. In temporal feature construction, the difference feature plays a key role and brings a great improvement in performance, but the variance feature does not show a performance improvement. In data imbalance alleviation, the under-sampling performed by k-means greatly loses the information of the healthy disks, and the over-sampling performed by SMOTE may synthesize failed disks with overlapping information, both of which impair the performance of the model. We believe that these analyses are significant in some extent to the future work in large-scale disk failure prediction.

Acknowledgment. This work is supported by National Natural Science Foundation of China under grants 61572408 and 61972326, and the grants of Xiamen University 20720180074.

References

1. Agarwal, V., Bhattacharyya, C., Niranjan, T., Susarla, S.: Discovering rules from disk events for predicting hard drive failures. In: Proceedings of the IEEE International Conference on Machine Learning and Applications, vol. 1, pp. 782–786. IEEE, December 2009
2. Botezatu, M.M., Giurgiu, I., Bogojeska, J., Wiesmann, D.: Predicting disk replacement towards reliable data centers. In: Proceedings of the ACM SIGKDD International Conference on Knowledge Discovery and Data Mining, vol. 13–17-August, pp. 39–48. ACM Press, August 2016
3. Chen, T., Guestrin, C.: Xgboost: a scalable tree boosting system. In: Proceedings of the ACM SIGKDD International Conference on Knowledge Discovery and Data Mining. vol. 13–17-August, pp. 785–794. ACM Press, August 2016
4. alibaba edu: The dataset of over 200 thousands hard disk drives in alibaba cloud's data centers (2020). https://github.com/alibaba-edu/dcbrain/tree/master/diskdata
5. Han, S., Lee, P.P., Shen, Z., He, C., Liu, Y., Huang, T.: Toward adaptive disk failure prediction via stream mining. In: Proceedings of the IEEE International Conference on Distributed Computing Systems (2020)
6. Kaur, K., Kaur, K.: Failure prediction and health status assessment of storage systems with decision trees. In: Luhach, A.K., Singh, D., Hsiung, P.-A., Hawari, K.B.G., Lingras, P., Singh, P.K. (eds.) ICAICR 2018. CCIS, vol. 955, pp. 366–376. Springer, Singapore (2019). https://doi.org/10.1007/978-981-13-3140-4_33
7. Kaur, K., Kaur, K.: Failure prediction, lead time estimation and health degree assessment for hard disk drives using voting based decision trees. Comput. Mater. Continua **60**(3), 913–946 (2019)
8. Ke, G., et al.: Lightgbm: A highly efficient gradient boosting decision tree. In: Proceedings of the International Conference on Neural Information Processing Systems, pp. 3149–3157. Curran Associates Inc. (2017)
9. Li, J., et al.: Hard drive failure prediction using classification and regression trees. In: Proceedings of Annual IEEE/IFIP International Conference on Dependable Systems and Networks, pp. 383–394. IEEE, June 2014
10. Li, J., Stones, R.J., Wang, G., Li, Z., Liu, X., Ding, J.: New metrics for disk failure prediction that go beyond prediction accuracy. IEEE Access **6**, 76627–76639 (2018)
11. Li, J., Stones, R.J., Wang, G., Li, Z., Liu, X., Xiao, K.: Being accurate is not enough: new metrics for disk failure prediction. In: Proceedings of IEEE International Symposium on Reliable Distributed Systems, pp. 71–80. IEEE, September 2016
12. Li, J., Stones, R.J., Wang, G., Liu, X., Li, Z., Xu, M.: Hard drive failure prediction using decision trees. Reliabil. Eng. Syst. Safety **164**, 55–65 (2017)

13. Mahdisoltani, F., Stefanovici, I., Schroeder, B.: Proactive error prediction to improve storage system reliability. In: Proceedings of the 2017 USENIX Annual Technical Conference, pp. 391–402. USENIX Association, July 2017
14. Murray, J.F., Hughes, G.F., Kreutz-Delgado, K.: Machine learning methods for predicting failures in hard drives: a multiple-instance application. J. Mach. Learn. Res. **6**, 783–816 (2005)
15. Pang, S., Jia, Y., Stones, R., Wang, G., Liu, X.: A combined bayesian network method for predicting drive failure times from smart attributes. In: International Joint Conference on Neural Networks, vol. 2016-October, pp. 4850–4856. IEEE, July 2016
16. Paris, J.F., Rincón, C.A.C., Vilalta, R., Cheng, A.M.K., Long, D.D.E.: Disk failure prediction in heterogeneous environments. In: Proceedings of the International Symposium on Performance Evaluation of Computer and Telecommunication Systems, pp. 1–7. IEEE, July 2017
17. Shen, J., Wan, J., Lim, S.J., Yu, L.: Random-forest-based failure prediction for hard disk drives. Int. J. Distrib. Sens. Netw. **14**(11), 155014771880648 (2018)
18. Wang, Y., Miao, Q., Pecht, M.: Health monitoring of hard disk drive based on mahalanobis distance. In: Proceedings of Prognostics and System Health Managment Confernece, pp. 1–8. IEEE, May 2011
19. Xiao, J., Xiong, Z., Wu, S., Yi, Y., Jin, H., Hu, K.: Disk failure prediction in data centers via online learning. In: Proceedings of the International Conference on Parallel Processing, pp. 1–10. ACM Press, August 2018
20. Xu, C., Wang, G., Liu, X., Guo, D., Liu, T.Y.: Health status assessment and failure prediction for hard drives with recurrent neural networks. IEEE Trans. Comput. **65**(11), 3502–3508 (2016)
21. Xu, Y., et al.: Improving service availability of cloud systems by predicting disk error. In: Proceedings of the 2018 USENIX Annual Technical Conference, pp. 481–494. USENIX Association, July 2018
22. Yang, W., Hu, D., Liu, Y., Wang, S., Jiang, T.: Hard drive failure prediction using big data. In: Proceedings of IEEE International Symposium on Reliable Distributed Systems Workshop, vol. 2016-January, pp. 13–18. IEEE, September 2015
23. Zhao, Y., Liu, X., Gan, S., Zheng, W.: Predicting disk failures with HMM- and HSMM-based approaches. In: Perner, P. (ed.) ICDM 2010. LNCS (LNAI), vol. 6171, pp. 390–404. Springer, Heidelberg (2010). https://doi.org/10.1007/978-3-642-14400-4_30
24. Zhu, B., Wang, G., Liu, X., Hu, D., Lin, S., Ma, J.: Proactive drive failure prediction for large scale storage systems. In: Proceedings of the IEEE Symposium on Mass Storage Systems and Technologies, pp. 1–5. IEEE, May 2013

SHARP: SMART HDD Anomaly Risk Prediction

Wei Liu, Yang Xue[✉], and Pan Liu

Institute for Infocomm Research, Singapore, Singapore
{liu_wei,yxue,liu_pan}@i2r.a-star.edu.sg

Abstract. With the fast expansion of online media and cloud-based storage, hard disk drive failure prediction becomes an increasingly important problem that has great industry impact. In the last 20 years, much effort has been put into using machine learning method to enhance the S.M.A.R.T monitoring system. Success has been achieved at various degrees, but the state-of-the-art methods still have considerable distance from the level of performance required by industry operations. In this paper, we demonstrated that with a strategic ensemble of models that cover both short-range and long-range temporal dependencies of S.M.A.R.T data, it is possible to achieve higher overall failure prediction accuracy and robustness. Our proposed model, named SHARP, is shown to achieve 56% F1 score in one of the holdout blind tests using an industry-scale data set. In the online competition test set, the F1 score was 38%.

Keywords: Hard disk drive · SMART · Failure prediction · Machine learning · Model ensemble

1 Introduction

Nowadays, with the surge of online media and digital content, data centers around the world are undergoing expansion at record speed. New hyper-scale data centers give rise to unprecedented demand on hard disks (HDD) [1]. With Petabytes of data being read and written at data centers on daily basis, the annual failure rate of HDD is estimated to be around 15% [2]. Given the large amount of HDD installed at data centers, which is in the range of hundreds of thousands, it is common for a data center to see tens of HDD failures occurring every single day [2]. HDD failures inevitably lead to data center's performance degradation, ranging from minor longer write time to serious catastrophic service breakdown [3]. In fact, it is found that most of the data center service problems are due to HDD failures [3].

Currently, most data centers manage HDD failures with the help of Self-Monitoring, Analysis and Reporting Technology (S.M.A.R.T). S.M.A.R.T was introduced in 1995 serving as standardized HDD failure warning system. By monitoring various hardware-related measurements, S.M.A.R.T detects signs that an HDD is likely to fail soon, typically within 24 h [4].

© Springer Nature Singapore Pte Ltd. 2020
C. He et al. (Eds.): AI Ops 2020, CCIS 1261, pp. 74–84, 2020.
https://doi.org/10.1007/978-981-15-7749-9_8

It is highly beneficial to have more lead time with HDD failure warnings, because it allows more time to replace the problematic HDD, or just to allow more conservative failure prevention measures. To achieve this end, data analysis and machine learning methods have been widely developed to make use of S.M.A.R.T data as well as other information relating to HDD operations to perform more accurate HDD failure predictions [4–21].

In [5], the authors demonstrated using naive Bayesian methods to perform anomaly detection and supervised learning. With expectation-maximization, a number of Naive Bayesian sub-models were trained and used for detecting abnormal HDD failures. To overcome noise in the data, binning method was used for smoothing. The authors experimented on an industrial data set and improved true positive rate from industry-standard 0.11 to 0.30. Although the improvement was apparent, the achieved accuracy was still not satisfactory enough for reliable prediction in industry setting.

In [15,20,21], the authors developed failure prediction models by formulating the failure prediction problem as a binary classification problem and solved using Classification and Regression Trees (CART) models [22]. The main advantage of CART-based method lies in its straightforward set-up, in which models are trained with isolated S.M.A.R.T data points and prediction can be made based on as few as one single S.M.A.R.T data point. This is especially helpful in the situation where only sporadic data are available during actual operation. In addition, recent CART models, e.g., XGBoost [23], has been shown to achieve good performance in HDD failure prediction problems. However, the CART-based method have to rely on hand-crafted time-window features to characterize any temporal data patterns, which are often not as effective as methods using sequential models, especially for long-range temporal dependency modeling.

S.M.A.R.T data of HDDs are time series in essence. It was observed that the HDD failures exhibited long-range temporal dependency with historical S.M.A.R.T data [6]. In other words, the eventual HDD failures were found to have higher correlation with a relatively long period of historical S.M.A.R.T data, e.g., in 30 days before failures as reported in [6], whereas the correlation between HDD failures and S.M.A.R.T data in only a day or two before failures were found to be lower.

Various work in the literature looked into the time-series nature of the data with the aim to further improve HDD failure prediction performance. In [7], the authors proposed a Hidden Markov Model (HMM) and Hidden Semi-Markov Model (HSMM)-based solution to predict HDD failures. The proposed solution was found to outperform benchmark methods, including Wilcoxon-Mann-Whitney rand-sum test on single-attribute data and Support Vector Machine on multi-attribute data. However, the proposed solution was susceptible to near-range temporal data noise; in addition, while HMM and HSMM are effective in modeling short-range dependencies, they are intrinsically not as efficient in modeling long-range temporal dependency [24], which renders them sub-ideal for HDD failure prediction applications. In [25], the authors employed Recurrent Neural Network (RNN) to model the long-range temporal dependency, and demonstrated improved HDD failure prediction performance. Instead of treating the health conditions of HDDs as either *good* or *bad*, the authors defined six grades of health conditions depending on the time until eventual failure. Better performance was obtained compared to Tree-based methods and multi-class Neural Network methods. The main drawback of the proposed method is its high requirement on training data. In order to train high quality RNN models, a large amount of high-quality training data, especially data associated with HDD failures, spanning a relative long range of past history are required. In some cases, such training data may not be available. In addition, HDDs may have gaps in data recording during operation, e.g., missing S.M.A.R.T data for a short period of time. The missing episodes of data may introduce noise to predictions made using RNN models.

In the past 20 years, while much effort has been dedicated to improving HDD failure prediction using various machine learning methods and models, it is clear that no single method or model is able to fully solve the problem. Partially due to the complex nature of HDD failure mechanisms that involve both long-range temporal dependency degradation failures and short-range temporal dependency catastrophic failures, which render it difficult for single model to fully characterize. To specifically address this challenge, we proposed SHARP (SMART HDD Anomaly Risk Prediction), an ensemble-based model that incorporate multiple classification models that target different HDD-failure-relevant S.M.A.R.T data patterns, which may relate to different failure mechanisms. We also explored the ensemble of CART models with sequential models to address long-range temporal dependency challenge.

The remainder of this paper is arranged as follows: in Sect. 2, we elaborate the methodology of SHARP; in Sect. 3, we demonstrate SHARP in experiments based on a public industry-scale data set; in Sect. 4, we summarize the major findings of our work and discuss future work.

2 Methodology

SHARP offers two modes to predict HDD failures. One is single-day-based prediction mode, the other is sequenced-day-based prediction mode.

2.1 Single-Day-Based Prediction Mode

In this mode, SHARP evaluates the risk probability of HDD failures solely base on each single day's S.M.A.R.T data of each HDD. The risk probability is a real number within the range of *0* and *1*, with *0* or *negative* means HDD failure is highly unlikely to happen in the next 30 days, and *1* or *positive* means HDD failure is almost certain to happen in the next 30 days. XGBoost is our preferred model for SHARP due to its robustness against highly imbalanced data. SHARP uses a 2-layer ensemble classifier as illustrated in Fig. 1. For Layer-1 classifier, SHARP takes as input of all S.M.A.R.T data, with *NA* excluded, and the rest of raw data Logarithmic-transformed.

The training data include two classes, namely positive class and negative class, where positive class include S.M.A.R.T data of all available failing HDDs on the day of failure, and negative class include S.M.A.R.T data of random selected healthy HDDs on random days subject to the condition that the sampled HDD would not fail in the next 90 days.

In the model training process, hyper-parameter tuning are performed in five-fold cross validation on six XGBoost hyper-parameters as follows,

- learning_rate
- n_estimators
- max_depth
- subsample
- reg_alpha
- reg_lambda

Subsequently, SHARP applies feature selection to achieve optimal accuracy in cross validation and retrain to obtains Layer-1 model. Layer-1 classifier is not expected to correctly detect all HDD failures, SHARP uses Layer-2 classifier to further improve its accuracy on HDD failures that escape Layer-1 classifier. The first step to train Layer-2 classifier is to obtain the training data. SHARP selects out-of-fold false negative HDD predictions in Layer-1 classifier's cross validation to be the positive class training data. The negative class is randomly re-sampled similar to Layer-1 negative class.

In terms of features, SHARP uses different features between Layer-1 and Layer-2 classifiers. While Layer-1 uses mostly Logarithmic-transformed features, Layer-2 is forced to use original normalized S.M.A.R.T data to further differentiate Layer-1 and Layer-2 classifiers. We need to force Layer-2 model to use only normalized S.M.A.R.T data because Layer-1 model has been using Logarithmic-transformed features, Layer-2 model should use different features to achieve different prediction results when using similar training instances.

For prediction process of a new HDD on a new day, which is illustrated in Fig. 2, SHARP takes as input of current day's S.M.A.R.T data of the HDD and pass them through Layer-1 models. Base on the threshold settings of Layer-1, SHARP predicts the HDD as positive or negative. If positive, the HDD will be recorded to Layer-1 True Prediction List. Else, HDD will be passed to Layer-2 models for another round of prediction. If Layer-2 model's prediction is positive, the HDD will be recorded to Layer-2 True Prediction List. Eventually, SHARP combines both Layer-1 True Prediction List and Layer-2 True Prediction List as its final prediction result.

The key in this mode is the Layer-2 classifier, which targets on false negative predictions from Layer-1 classifier and uses different features to construct the model. Layer-2 model helps SHARP to successfully predict more HDD failures, but the single-day-based prediction mode is still not sensitive enough to failures of slow degradation nature, which we attempt to address using sequential models.

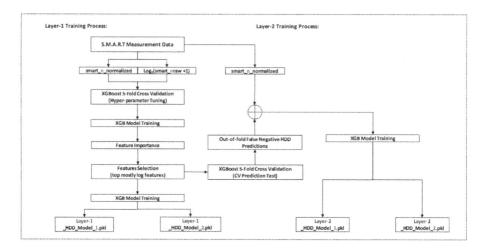

Fig. 1. SHARP uses a two layer classifier ensemble to increase the sensitivity in predicting HDD failures of different data patterns

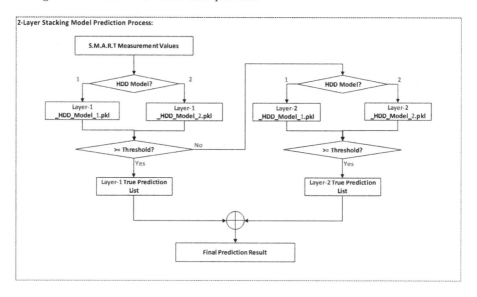

Fig. 2. SHARP makes prediction based on the combined prediction results of Layer-1 and Layer-2 classifiers

2.2   Sequenced-Day-Base Prediction Approach

The purpose of this approach is to pick-up those HDD failures with slow degradation and to extend prediction lead time. The S.M.A.R.T measurement data are time-series data recorded everyday. Both the daily performance and the performance deviation from previous days may be important in correctly predicting HDD failures. SHARP uses Gated Recurrent Unit (GRU) models to perform sequential modeling for HDDs failure with long-range temporal dependency.

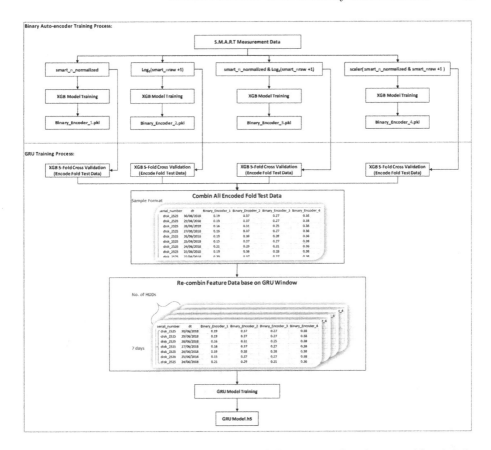

Fig. 3. GRU models the long-range temporal dependency by the ensemble of daily prediction probabilities of multiple XGBoost classifiers

As shown in Fig. 3, to perform GRU ensemble, SHARP uses four independently trained XGBoost binary classifiers to encode daily S.M.A.R.T performance. The performance rating is, similar to single-day-based prediction mode, from *0* to *1*. To ensure model diversity for ensemble performance, each XGBoost model should be trained with different features, and XGBoost hyperparameters. GRU, which is then trained based on prediction probability of individual XGBoost classifiers, essentially selects the useful encoded values for modelling according to input data.

With encoded feature, we can easily visualize S.M.A.R.T performance by the day (Fig. 4). By visualizing failure cases, we can identify the optimal number to use as GRU monitoring window. GRU training data is generated from XGBoost cross validation by first encoding test data in each fold, then combining all binary auto-encoder output into one and re-organizing them by HDD series number.

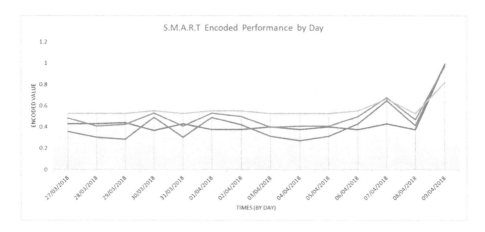

Fig. 4. GRU models enables easy visualization of HDD health condition by the day

GRU prediction is straightforward (see Fig. 5). GRU model prediction need handle multiple days data at the same time. Compare to 2-Layer Stacking approach, GRU model prediction require more computation power. Binary encoders will encode these 7 days S.M.A.R.T measurement into 4 columns. We need reshape those 4 columns data by number_of_HDDs × 7 × 4. If any of HDD is missing data in these 7 days, the leading day point will be used for padding. For GRU final decision, we use threshold control to further boost the accuracy. The threshold can be learn from test data-set.

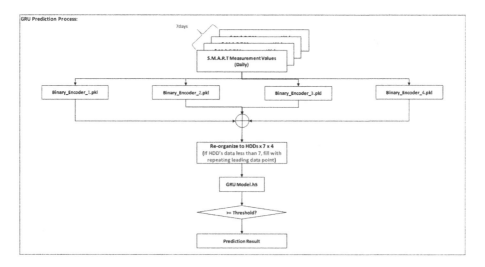

Fig. 5. GRU Model Prediction Process looks at past S.M.A.R.T of a few days to predict HDD failures

The key in this mode is encoder output as the input of GRU. The S.M.A.R.T measurement is lack of connection between HDDs. This reason is limiting us to use DNN and CNN for auto-encoding network. We indeed try to use 1D-CNN for auto-encoding training, the result is not promising. In the end, we decided to use binary auto-encoder which is not rely on the connection between HDDs.

3   Experiment

To evaluate the performance of SHARP under these two modes, we use Alibaba S.M.A.R.T data set [26]. This data set consists of two types of files, disk_sample_smart_log_*.csv and disk_sample_fault_tag.csv.

The disk_sample_smart_log_*.csv is the daily S.M.A.R.T data of traditional HDD which has 514 columns. Most data are empty, only 48 measurements are usable for training model. Details as Table 1. The 'disk_sample_fault_tag.csv' file is the fault disk labels. The experiment uses data set range from 2017-07-31 to 2018-06-30 for training models in both approaches and uses data set range from 2018-07-01 to 2018-07-31 for evaluating their performance. For preparing this evaluation data set, we use all failure HDDs and 10,000 good HDDs (random selection) to consist this evaluation data set.

Table 1. S.M.A.R.T log data information.

Field	Type	Number of columns	Description
serial_number	String	1	Disk serial number code
manufacturer	String	1	Disk manufacturer code
model	String	1	Disk model code
smart_n_normalized	Integer	24	Normalized SMART data of SMART ID=n
smart_nraw	Integer	24	Raw SMART data of SMART ID=n
dt	String	1	Sampling time in format (yyyy-mm-dd)

Experiment results can be seen in Fig. 6. Base on evaluation data set, Layer-1 of 2-Layer ensemble model is able to reach the highest F1 score 56% by using threshold 0.95. With additional Layer-2, the overall F1 score is able to remain the same value. But the recall is increased from 51.3% to 55.8%. This result is tally with our design purpose in the second layer. The GRU F1 score 49% is lower than 2-Layer Stacking Model when using threshold 0.9, but the overall F1 score is much stable than 2-Layer ensemble model. The 2-Layer Stacking Model is rely on threshold, and these thresholds may bias to certain data-sets. This result indicates that GRU could be relatively more robust. In general, neural network requires much more data than machine learning model. In Alibaba data-set, the quantity of failure records is about 1% of good records only. Due to the relatively low quantity of training data, the GRU performance is limited.

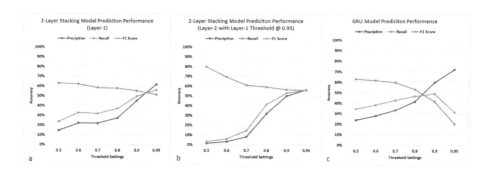

Fig. 6. Experiment result

4   Conclusion

We presented SHARP, a machine learning based HDD failure prediction model. In experiments on industry-scale data, SHARP is shown to achieve high and stable performance over different data sets. In holdout blind test, it achieved F1 score of 56% and 49% respectively in single-day-based prediction mode and sequenced-day-based prediction mode. It is demonstrated by the ensemble of multiple sufficiently-different models, it is possible to achieve higher prediction performance. In addition, by the ensemble of a group of single-day-based models with sequential models, e.g., GRU, it is demonstrated that a more robust prediction accuracy over long-range dependency HDD failures could be achieved.

During our work, it is noted that input data of GRU models should be simple and focused to achieve stable performance. Complex and high dimensional data will request large quantity of high quality data in GRU training. Auto-encoder is a useful technique to convert a complex and high dimensional data into simple and focused features. As part of the future work, we plan to explore various encoding method, e.g., Convolutional Neural Network-based auto-encoder for feature engineering.

References

1. Miller, R.: Google building more data centers for massive future clouds. https://datacenterfrontier.com/google-building-more-data-centers-for-massive-future-clouds

2. Huang, S., Liang, S., Fu, S., Shi, W., Tiwari, D., Chen, H.: Characterizing disk health degradation and proactively protecting against disk failures for reliable storage systems. In 2019 IEEE International Conference on Autonomic Computing (ICAC), pp. 157–166. IEEE (2019)
3. Ganguly, S., Consul, A., Khan, A., Bussone, B., Richards, J., Miguel, A.: A practical approach to hard disk failure prediction in cloud platforms: big data model for failure management in datacenters. In 2016 IEEE Second International Conference on Big Data Computing Service and Applications (BigDataService), pp. 105–116. IEEE (2016)
4. Hughes, G.F., Murray, J.F., Kreutz-Delgado, K., Elkan, C.: Improved disk-drive failure warnings. IEEE Trans. Reliab. **51**(3), 350–357 (2002)
5. Hamerly, Greg, Elkan, Charles, et al.: Bayesian approaches to failure prediction for disk drives. In: ICML, vol. 1, pp. 202–209 (2001)
6. Chang, X., Wang, G., Liu, X., Guo, D., Liu, T.-Y.: Health status assessment and failure prediction for hard drives with recurrent neural networks. IEEE Trans. Comput. **65**(11), 3502–3508 (2016)
7. Zhao, Y., Liu, X., Gan, S., Zheng, W.: Predicting disk failures with HMM- and HSMM-based approaches. In: Perner, P. (ed.) ICDM 2010. LNCS (LNAI), vol. 6171, pp. 390–404. Springer, Heidelberg (2010). https://doi.org/10.1007/978-3-642-14400-4_30
8. Wang, Y., Miao, Q., Pecht, M.: Health monitoring of hard disk drive based on Mahalanobis distance. In: 2011 Prognostics and System Health Management Conference, pp. 1–8. IEEE (2011)
9. Zhu, B., Wang, G., Liu, X., Hu, D., Lin, S., Ma, J.: Proactive drive failure prediction for large scale storage systems. In: 2013 IEEE 29th Symposium on Mass Storage Systems and Technologies (MSST), pp. 1–5. IEEE (2013)
10. Li, J., et al.: Hard drive failure prediction using classification and regression trees. In: 2014 44th Annual IEEE/IFIP International Conference on Dependable Systems and Networks, pp. 383–394. IEEE (2014)
11. Yang, W., Hu, D., Liu, Y., Wang, S., Jiang, T.: Hard drive failure prediction using big data. In: 2015 IEEE 34th Symposium on Reliable Distributed Systems Workshop (SRDSW), pp. 13–18. IEEE (2015)
12. Pang, S., Jia, Y., Stones, R., Wang, G., Liu, X.: A combined Bayesian network method for predicting drive failure times from smart attributes. In: 2016 International Joint Conference on Neural Networks (IJCNN), pp. 4850–4856. IEEE (2016)
13. Botezatu, M.M., Giurgiu, I., Bogojeska, J., Wiesmann, D.: Predicting disk replacement towards reliable data centers. In: Proceedings of the 22nd ACM SIGKDD International Conference on Knowledge Discovery and Data Mining, pp. 39–48 (2016)
14. Rincón, C.A.C., Pâris, J.-F., Vilalta, R., Cheng, A.M.K., Long, D.D.E.: Disk failure prediction in heterogeneous environments. In: 2017 International Symposium on Performance Evaluation of Computer and Telecommunication Systems (SPECTS), pp. 1–7. IEEE (2017)
15. Li, J., Stones, R.J., Wang, G., Liu, X., Li, Z., Xu, M.: Hard drive failure prediction using decision trees. Reliab. Eng. Syst. Saf. **164**, 55–65 (2017)
16. Mahdisoltani, F., Stefanovici, I., Schroeder, B.: Proactive error prediction to improve storage system reliability. In: 2017 {USENIX} Annual Technical Conference ({USENIX}{ATC} 2017), pp. 391–402 (2017)
17. Xiao, J., Xiong, Z., Wu, S., Yi, Y., Jin, H., Hu, K.: Disk failure prediction in data centers via online learning. In: Proceedings of the 47th International Conference on Parallel Processing, pp. 1–10 (2018)

18. Xu, Y., et al.: Improving service availability of cloud systems by predicting disk error. In: 2018 {USENIX} Annual Technical Conference ({USENIX}{ATC} 2018), pp. 481–494 (2018)
19. Shen, J., Wan, J., Lim, S.-J., Lifeng, Yu.: Random-forest-based failure prediction for hard disk drives. Int. J. Distrib. Sensor Netw. **14**(11), 1550147718806480 (2018)
20. Kaur, K., Kaur, K.: Failure prediction and health status assessment of storage systems with decision trees. In: Luhach, A.K., et al. (eds.) ICAICR 2018. CCIS, vol. 955, pp. 366–376. Springer, Singapore (2019). https://doi.org/10.1007/978-981-13-3140-4_33
21. Kaur, K., Kaur, K.: Failure prediction, lead time estimation and health degree assessment for hard disk drives using voting based decision trees. CMC Comput. Mater. Continua **60**, 913–946 (2019)
22. Breiman, L., Friedman, J.H., Olshen, R., Stone, C.J.: Classification and Regression Trees. CRC Press, Boca Raton (1984)
23. Chen, T., Guestrin, C.: XGBoost: a scalable tree boosting system. In: Proceedings of the 22nd ACM SIGKDD International Conference on Knowledge Discovery and Data Mining, pp. 785–794 (2016)
24. Yoon, B.-J., Vaidyanathan, P.P.: Context-sensitive hidden Markov models for modeling long-range dependencies in symbol sequences. IEEE Trans. Signal Process. **54**(11), 4169–4184 (2006)
25. Wang, Y., Jiang, S., He, L., Peng, Y., Chow, T.W.S.: Hard disk drives failure detection using a dynamic tracking method. In: 2019 IEEE 17th International Conference on Industrial Informatics (INDIN), vol. 1, pp. 1473–1477. IEEE (2019)
26. Alibaba S.M.A.R.T data-set. https://github.com/alibaba-edu/dcbrain/tree/master/diskdata

Tree-Based Model with Advanced Data Preprocessing for Large Scale Hard Disk Failure Prediction

Qi Wu[1], Weilong Chen[2], Wei Bao[3], Jipeng Li[2], Peikai Pan[1], Qiyao Peng[1], and Pengfei Jiao[1(✉)]

[1] Tianjin University, Tianjin, China
Morxrc@163.com, {ppan,qypeng,pjiao}@tju.edu.cn
[2] University of Electronic Science and Technology, Chengdu, China
chenweilongg921@gmail.com, jipeng416@gmail.com
[3] Southeast University, Nanjing, China
willinseu@gmail.com

Abstract. As the scale of data in data centers expands, the hard drives are widely used in computer. However, hard disk failures occur frequently in actual scenarios. With the increase of utilizing time, the stability and accuracy of hard disk are continuously decreasing, and will result in negative impact on normal operation of the system. However, there are no researches on the estimation of hard disk quality in entire industry. In this article, we utilize Generative Adversarial Networks (GAN) for realizing data augmentation, and use the catboost model to model the prediction of disk damage, which achieved tenth place in the PAKDD2020 Alibaba intelligent operation and maintenance algorithm competition-large-scale hard disk failure prediction competition [1].

Keywords: Failure prediction · Hard disk drives · Classification

1 Introduction

Hard disk drives (HDDs) are not only among the most frequently failing components in computer today, but also the main reasons in server failures [2]. It has been estimated that HDDs faults caused by lots of unprecedented storage systems account for 78% of the hardware replacements in Internet data centers (IDCs) [3]. The consequences of HDDs failures might be permanent and difficult to be recovered, even unrecoverable, which lead to longer server downtime and lower reliability of IDCs [4]. The forecasting of HDDs failure is difficult, which aims at predicting the possible failures of HDDs in advance for making the storage system more stable and reliable. The basic strategy of HDDs prediction is that, if the impending disk failures have been detected or predicted, users can be informed to take measures such as backup data in advance to decrease the losses cased by HDDs accident.

© Springer Nature Singapore Pte Ltd. 2020
C. He et al. (Eds.): AI Ops 2020, CCIS 1261, pp. 85–99, 2020.
https://doi.org/10.1007/978-981-15-7749-9_9

Nowadays, most of HDDs have been equipped with Self-Monitoring, Analysis and Reporting Technology (SMART) [5], which are implemented to evaluate drivers health status on the inner condition and environments data of HDDs provided by sensors and counter [6]. If the situation of HDDs failures was detected by this technology, the administrator would be informed. However, the original methodology might be not get a high prediction accuracy, which ranges from 3% to 10%, And 0.1% of the failure alarms are misstatements [7]. It can't be denied that although the SMART performs not satisfactory, the data and attributes collected by this technology could be useful to detect HDDs failures exceedingly.

There have been several feasible improvements to improve the accuracy of HDDs failure prediction and decrease the error rate. Several methods are based on machine learning, deep learning and other statistical approaches, which regarded the SMART attributes as input [8–11]. We adopt machine learning method CatBoost [30] proposed by Liudmila Prokhorenkova et al. To construct the key model in the competition, which consist of catgorical and boost. This model can solve the problem of gradient bias and prediction shift, which can significantly improve the model's classification accuracy and generalization ability. In order to facilitate the training process, we extract 100-dimensional features by utilizing the sampled data of July. Then, we use ROZ (remove-one-zero) method to maintain model's stability against mutation data. Despite there are various existing methodology in this field, applying machine learning to predict HDDs failure still faces two practical challenges:

1) One challenge is that the dataset collected by SMART is not complete. For instance, previous researched reported that more than half of SMART failure signals are missing in failed disks [12]. HDDs failure prediction won't be accurate if the dataset is incomplete. HDDs monitoring systems may stop recording failure signals due to network failures, software maintenance/upgrades, system crashes, and human mistakes in production [13]. Some special commercial activities may require the suspension of the disk monitoring systems for server offloading [14]. One solution is to interpolate the missing failure signals. We adopt the preprocessing technique spline-based data filling, which fills the values of missing samples via cubic spline interpolation [15] to account for any possible abrupt changes in such missing samples.

2) The other challenge is that the training data is imbalanced. Although some approaches [8,11,16] have performed well in HDDs failure prediction, they suffer from the data imbalance issue heavily, i.e., the amount of healthy disks is much larger than that of failed ones. Nevertheless, it is very important for classifier training process to use balanced dataset [17]. So the data imbalance issue may greatly decrease the accuracy of HDDs failure prediction. What is worse, the training data is gradually gathered instead of being given in advance [8]. As a result, the training data collected within the initial period may be insufficient and could result in an inability of the predictor at the beginning of its deployment, i.e., cold starting problem [5]. To overcome this challenge, we consider using Generative Adversarial Networks (GANs), which is able to capture the data distribution and deal with data imbalance.

To sum up, the contributions of our method are summerized as follows:

- We select the CatBoost algorithm to model and extract 100-dimensional feature from the sampled data of July. The model performs really stable and efficient no matter what kind of data are imported into the model.
- We use ROZ (remove-one-zero) method, which has very strong stability against mutation data. The experimental results demonstrate the effectiveness of our method.
- In order to solve the problem of data missing, we adopt spline-based data filling method to fill the missing values, which are perfectly fit to the existing data.
- We utilize GAN as the data augmentation method and the model becomes extremely robust.

The rest of this paper is organized as follows: we first introduce the background of HDDs failure prediction and existing work in Sect. 2. Following it, Sect. 3 presents introduced our prediction model and relevant algorithms and technologies, including CatBoost, data preprocessing methodology, generative adversarial networks, feature engineering, et al. and Sect. 4 introduces the experiments. Finally, Sect. 5 concludes this paper.

2 Related Work

The methods of predicting disk failure based on SMART attributes are mainly divided into statistical approaches and machine learning approaches.

Statistical Approaches. Statistical approaches mostly include rank-sum test and Bayesian approaches. (1) Rank-sum test approaches: Hughes et al. [18] found that many SMART attributes are non-parametrically distributed and applied a multivariate rank-sum test and OR-ed single variate test to 3744 drives in which 36 drives had failed. The result of the experiment achieved an FDR (failure detection rate) of 60% with an FAR (false alarm rate) of 0.5%. Murray et al. [19] compared the performance of SVM, unsupervised clustering, rank-sum, and reverse arrangements tests. The results showed that the rank-sum test obtained the best performance: an FDR of 33.2% with an FAR of 0.5%. The experimental dataset included 369 hard drives, 178 of which are good drives and 191 of which are failed drives.

(2) Bayesian approaches: Hamerly and Elkan [20] employed two Bayesian approaches, naive-Bayes expectation-maximization (NBEM) and a semi-supervised method, to predict drive failure. The experimental dataset includes 1927 good drives and 9 failed drives. These two approaches achieved FDRs of 35%–55% with FARs of approximately 1%. Then, Murray et al. [21] used the same dataset and proposed an algorithm based on the multiple-instance learning framework and the naive Bayesian classifier. They found that SVM using all 25 attributes achieved the best prediction performance, with an FDR of 50.6% and an FAR of 0%; however, rank-sum test outperformed SVM for the small part of

SMART attributes. Ma et al. [22] proposed RAIDShield to predict drive failures on RAID storage systems. RAIDShield used the conditional possibility of RSC and Bayes to predict the RAID failures. The experimental data were from data backup systems at EMC Inc. RAIDShield eliminated 88% of triple disk errors.

Machine Learning Approaches. Machine learning approaches applied to failure prediction include SVM, BPNN,classification and regression tree (CART) and so on. Zhu et al. [16] developed a BPNN model and an improved SVM model on a SMART dataset, containing 22,962 good drives and 433 failed drives. The SVM model achieves the lowest FAR (0.03%), and the BP neural network model is far superior in detection rate which is more than 95% while keeping a reasonable low FAR. Qian et al. [23] proposed a priority-based proactive prediction (P3) algorithm and got evaluation results of 86.3% prediction rate and 0.52% false alarm rate. However, the BPNN did not achieve satisfactory performance stability and interpretability.

Wang et al. [24] developed an approach for HDD anomaly detection using Mahalanobis distance (MD). Furthermore, Wang et al. [25] proposed a two-step parametric (TSP) method to achieve an FDR of 68.4% with an FAR of 0%. TSP method detected anomalies first, then used a sliding-window-based generalized likelihood ratio test to track the anomaly progression. They used failure modes, mechanisms, and effects analysis (FMMEA) 18 to select features and the minimum redundancy maximum relevance to remove the redundant features. Queiroz et al. [26] proposed an HDD fault detection method based on a combination of semi-parametric and nonparametric models to overcome the limitation of distributions of the SMART attributes.

Xu et al. [27] proposed an RNN-based model for health status assessment and failure prediction for HDDs. The RNN-based model achieved a high-prediction performance, with FDRs of 87%–97.7% and FARs of 0.004%–0.59%. Jiang et al. [5] presented SPA, a GAN-based anomaly detection approach, to predict lifelong disk failure. The model is trained end-to-end by leveraging CNN's feature extraction characteristic which captures the temporal locality contained in constructed image-like 2D-SMART attributes.

Li et al. [9] proposed two prediction models based on CT and CART, respectively, and utilized the health degree to describe the deterioration process. The health degree model was determined by the size of deterioration window and the number of hours before failure. The experimental dataset was from the data center of Baidu, containing 25,792 drives of 3 models. They achieved a high-prediction performance, with an FDR of 95% and an FAR of 0.1%. Shen et al. [4] proposed a method based on the part-voting random forest to improve the detection accuracy of soon-to-fail HDDs. The method differentiates prediction of HDD failures in a coarse-grained manner by part-voting and similarity between health samples, and achieved good performance: an FDR of 97.67% with an FAR of 0.017% for family 'B'; an FDR of 100% with an FAR of 1.764% for family 'S'; and an FDR of 94.89% with an FAR of 0.44% for family 'T'.

Lu et al. [28] used a large-scale dataset, including SMART attributes, performance metrics and location markers, to trained neural network models and the model predict disk failures with 0.95 F-measure and 0.95 MCC for 10 days prediction horizon (Table 1).

Table 1. The table compares part of related work of disk failure prediction, which the first half is statistical approaches and the second half is machine learning approaches. Column 'dataset' represents: number of disk devices (number of failure disk devices).

Related work	FDR	FAR	dataset
Hughes et al. (2002)	60%	0.5%	3,744 (36)
Murray et al. (2003)	33.2%	0.5%	369 (191)
Hamerly and Elkan (2001)	35%–55%	1%	1,936 (9)
Murray et al. (2005)	50.6%	0%	1,936 (9)
Zhu et al. (2013)	95%+	0.03%	23,395 (433)
Qian et al. (2015)	86.3%	0.52%	7,148 (130)
Wang et al. (2014)	68.4%	0%	369 (191)
Li et al. (2014)	95%	0.1%	25,792
Xu et al. (2015)	87%–97.7%	0.004%–0.59%	25,792
Shen et al. (2018)	94.89%–100%	0.017%–0.44%	75,428

In this paper, we transform the disk failure prediction into an anomaly detection problem and propose a novel tree-based method. We use the ways of ROZ, RDF and Smart-GAN to process data to make the model has stronger robustness and higher accuracy in disk failure prediction. Our method got the 10th place in the PAKDD2020 Alibaba intelligent operation and maintenance algorithm competition-large-scalehard disk failure prediction competition, which proved the method we proposed is effective.

3 Methodology

In this section, we will introduce our Methodology.

3.1 Data Preprocessing

We proposed a method of data preprocessing: ROZ (Remove-One-Zero). From the reality, we know that the disks do not fail immediately. With a long time small errors accumulating, the disks fail. However, when we use the binary classification, there would be inevitable mutations. If we directly use the history data of disks which have failure and ignore the mutations between two consecutive time points, the model will get confused with the positive data and negative data. The Fig. 1 shows that when disks fail, there SMART attributes may not change.

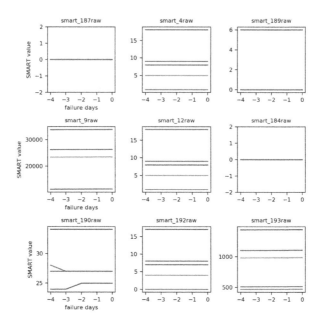

Fig. 1. This figure shows that even when the disk get a failure, the value of the smart do not change rapidly

---

Algorithm 1 Remove-One-Zero

Input: Historical data of disks D
Output: Selected historical data of disks D_{train}
1: Load only last day data of the fail disks D_{positive}
2: Load historical data of disks which do not fail D_{negative}
3: Concatenate two data D_{negative} and D_{positive} and train a model $Model_{\text{ROZ}}$
4: Load historical data of disks which fail in the last day $D_{\text{positive\_history}}$
5: set a *threshold*
6: **for** $i = 0$ to length $(D_{\text{positive\_history}})$ **do**
7:    $p_i = Model_{\text{ROZ}} \left(D^i_{\text{positive\_history}} \right)$
8:    **if** $p_i > threshold$ **then**
9:       $\text{label}_{D^i_{positive\_history}} = 1$
10:   **else**
11:      $\text{label}_{D^i_{positive\_history}} = 0$
12:   **end if**
13: **end for**
14: Concatenate $D'_{\text{positive\_history}}$ whose label $= 1$ and D_{negative} as D'
15: Output D'

Considering this, we have to remove the mutations which may hurt the model performance. We use the Remove-One-Zero method, which means we only use the history data which show the obvious failure. Firstly, we define the day of failure as the positive label, and other disks which do not have failures as the negative label. We train a model to define the 'failure' and use our model to give the unlabeled data a label near the day of failure, which contains positive label and negative label. After doing this, we remove all the negative part whose disks have failures. Finally, we train the online predict model with negative data whose disk do not fail and the positive data whose disk fail. Algorithm 1 describes the overall procedure for preprocessing and by using this procedure, we can get rid of the mutations which may mislead the model.

3.2 T-Valid

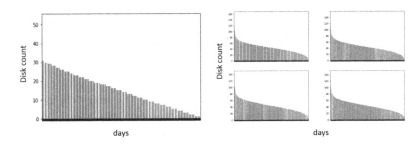

Fig. 2. The distribution of the online data (left) and T-valid data (right)

We proposed a new way of validation: T-Valid. As we can see from the Fig. 2, the distribution of the online data shows that as time pass by, the amount of data slowly decreases. It can be explained that if the disks fail, the disks can not be able to provide more data. Before the model goes online, we have to know whether the model has a good performance by using the validation. When we train the model, the validation can measure our model's performance. However, if the validation can not reflect the true improvement, we do not know whether our methods work. A verification set that accurately reflects the true performance of the model is particularly important.

According to this, we use the T-Valid method. The Fig. 3 shows the distribution of the T-Valid validation set. We truncate the original valid data to fit the online distribution. First, we leave the last 30 days of data in the training set as our original validation set. Then, according to the distribution online, we sample the daily data. While sampling, we guarantee that future data will appear in the current. By this procedure for preprocessing, our offline verification set matches the online test set well.

3.3 Data Preprocessing-SDF

Ideally, the system will gather smart data of all disks every day. But in real life, it is often unsatisfactory due to system failure or aging of disks. We found that there are many missing values in the dataset, which have a serious impact on our model and features. Therefore, before feature engineering, we must first fill in these missing values in every failed or healthy disk. We will first discuss several different filling methods.

Data Filling Approaches. The most basic way is to use forward filling, which means fill missing values with previous data (e.g., "100, 110, miss, 120, 130". The forward filling method will fill the missing value with 110). Another method is linear interpolation. The linear interpolation is a method of curve fitting using linear polynomials to construct new data points within the range of a discrete set of known data points (e.g., for the series "100, 110, miss, 120, 130"). The linear interpolation method will fill the missing value with 115). However, although the above two methods are very easy to understand and easy to implement, because the disk failure prediction problem is essentially an exception detection problem, and smart attributes often change dramatically, and the above missing value filling method can not capture this change very well.

Spline-Based Data Filling. In order to cope with the sharp changes caused by abnormal values, we use Cubic Spline Interpolation method [15] to fill the missing value. Spline interpolation is based on cubic spline, which is a spline constructed of piecewise third-order polynomials which pass through a set of m control points. Experimental results show that the cubic spline interpolation guarantees the smooth filling of missing values and gives good results to our model. However, cubic spline interpolation can not deal with the missing values of the beginning and the end. So, We directly delete the missing values at the beginning and end.

3.4   Smart-GAN

Generative Adversarial Networks [29] (GAN) is a deep learning model and one of the most promising methods for unsupervised learning on complex distributions in recent years. The model generates a fairly good output through the mutual game learning of (at least) two modules in the framework: the generative model and the discriminant model. In this work, in order to expand the sample size, we use GAN to generate fake samples with similar features. The F1 score of the model trained on the expanded data set can be increased by 5 thousandths.

When learning the distribution of a given data set, generative adversarial networks (GAN) have shown strong versatility. The basic GAN optimization process consists of two interacting networks. The first type is called a generator, which uses random vectors as input and generates sample distributions that are closer to the real data set as possible. The second one is called the discriminator trying to distinguish the real data set from the generated samples. In convergence, ideally, it is expected that the generator generates samples with the same distribution as the real data set.

To be specific, in a conventional GAN, G is trained to map a random noise vector $z \sim p_z$ (which is the distribution of the random noise) to the data space,

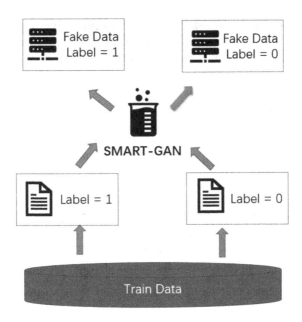

Fig. 3. Smart-GAN flow in our work

with the objective to maximize the probability that D classifies the samples from G as "real" data:

$$\mathbb{E}_{z \sim p_z}[\log D(G(Z))] \tag{1}$$

In contrast, D is trained to maximize the probability of assigning correct labels to the samples:

$$\mathbb{E}_{y \sim p_d}[\log D(y)] + \mathbb{E}_{z \sim p_z}[\log 1 - D(G(Z))] \tag{2}$$

where pd is the distribution of the dataset. Detailed structure is shown in Fig. 4.

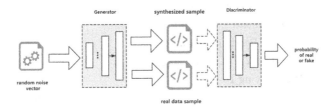

Fig. 4. Structure of smart-GAN

However, for our data augmentation problem, considering that the essence of our problem is a binary classification problem with supervised learning, we first divide the data into two categories according to the label, and the classification criterion is whether the hard disk damage will occur in the next 30 days. Then GAN is applied to the positive sample and the negative sample respectively for data expansion, the expansion ratio is 1:2 and 1:1 respectively. The process can be seen in Fig. 3.

3.5   Feature Engineer

For each disk data, it contains many smart attributes, including a raw value and a normalized value for each smart attribute. Normalized values are obtained by mapping the related raw value to one byte using vendor-specific methods. Because there is a certain correspondence between the raw values and the normalized values, we only keep the raw value column in the smart attributes. Then, we delete columns with all null values. We build a Lightgbm model to get the feature importance. According to the feature importance and a manually set threshold, the selected top 18 features are shown in Table 2.

Table 2. The 18 selected SMART attributes

Attribute ID	Attribute name	Attribute type
1	Read Error Rate	Raw
4	Start/Stop Count	Raw
5	Reallocated Sectors Count	Raw
7	Seek Error Rate	Raw
9	Power-On Hours	Raw
12	Power Cycle Count	Raw
184	End-to-End error	Raw
187	Reported Uncorrectable Errors	Raw
188	Command Timeout	Raw
189	High Fly Writes	Raw
190	Temperature Difference or Airflow Temperature	Raw
192	Power-off Retract Count	Raw
193	Load/Unload Cycle Count	Raw
197	Current Pending Sector Count	Raw
199	UltraDMA CRC Error Count	Raw
240	Head Flying Hours	Raw
241	Total LBAs Written	Raw
242	Total LBAs Read	Raw

In order to avoid the large number of features generated by feature engineering, We divide features into three groups.

- **1. Count type columns**: Count type columns include features such as SM4, SM5, SM12, SM187 and SM197, we first calculate the difference values over two consecutive samples in the time-series. Then we calculate the statistical characteristics of the difference feature according to the data of a window: mean, median, max, min and ptp (the difference between the maximum value and the minimum value in the object). We use Three window sizes,5,10 and 20 days.
- **2. Rate type columns**: We use the window 5, 10 and 20 days to extract statistical features from numerical features: standard deviation, median, mean, max, min.
- **3. Other columns**: For the other columns, we directly make a difference with the average of the past 20 days.

In addition to the above statistical characteristics, we find that the continuous growth, decline, or long-term constant of some SMART features may represent that the disk is in different states. Therefore, we respectively count the continuous growth days of SM5, SM197, the continuous decline days of SM1, SM7, SM184 and the stable state days of SM5, SM7, SM184, SM197 and SM198.

3.6 Catboost

We use Catboost [30] as our model algorithm. Catboost is an open-source machine learning library of Russian search giant yandex in 2017, which is a kind of boosting family algorithm. Catboost, XGboost and LightGBM are all improved implementations under the framework of GBDT algorithm. Catboost is a GBDT framework based on symmetric decision tree, which has fewer parameters, supports category variables and high accuracy. The main problem is to deal with category features efficiently and reasonably. Moreover, Catboost also solves the problems of gradient bias and prediction shift, so as to reduce the occurrence of over fitting and improve the accuracy and generalization ability of the algorithm. Compared with XGboost and LightGBM, Catboost has the following innovations:

- Using complete symmetric tree as base model.
- An innovative algorithm is embedded to automatically process the category features into numerical ones. Firstly, the algorithm make some statistics on the category features and calculate the frequency of a category feature, and then add the super parameters to generate new numerical features.
- Catboost also uses combined category features, which can make use of the relationship between features and enrich feature dimensions.
- In order to avoid the bias of gradient estimation and solve the problem of prediction bias, the method of sequence lifting is used to combat the noise points in the training set.

3.7   Ensemble

For the ensemble we adopt the strategy of boosting so as to decrease bias. We construct a strong learner by averaging the outcome of CatBoost meta-learner, which are selected in random. The combination of the weak learner are shown in Fig. 5.

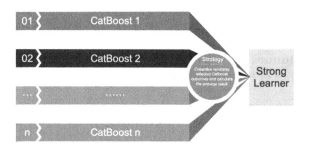

Fig. 5. The ensemble process

4   Experimental Results

We randomly sampled 22723 hard disks from 183033 of the original dataset. With only 442 hard disks failed and the rest of disks (22281) in the good class. To evaluate the performance of all methods we use T-valid to compute two metrics: the Area Under the ROC Curve (AUC) and the F1-score (showed in Table 3).

Table 3. Experimental results

Method	AUC	F1 score
Catboost	0.69742	0.2931
Catboost+ROZ	0.72165	0.3186
Catboost+SDF	0.72983	0.3257
Catboost+ROZ+SDF	0.73421	0.3214
Catboost+Smart-Gan	0.74865	0.3402
Catboost+ROZ+Smart-Gan	0.7492	0.3551
Catboost+SDF+Smart-Gan	0.7543	0.3570
Catboost+ROZ+SDF+Smart-Gan	0.76577	0.3818

5   Conclusion and Future Work

We propose a novel tree-based anomaly detection approach for lifelong disk failure prediction. By the data preprocessing ways including ROZ and SDF, we make our model more robust and higher accuracy in the disk failure prediction. We also propose the data enhancement way named Smart-GAN which adds more data during the training time, and kindly solves the unbalanced problem. Furthermore, we do some divided group feature engineering, avoiding the large number of features and accelerating the computing. We achieved tenth place in the PAKDD2020 Alibaba intelligent operation and maintenance algorithm competition-large-scalehard disk failure prediction competition. The results confirm that the proposed approach is effective. We will have more work on the reinforcement learning for the disk failure prediction and try to optimize long-term rewards, which may work better on this prediction.

Acknowledgments. This work was supported by the National Key R&D Program of China (2018YFC08 32103, 2018YFC0831000, 2018YFC0832101) and National Social Science Foundation of China (15BGL035).

References

1. https://github.com/alibaba-edu/dcbrain/tree/master/diskdata
2. Botezatu, M.M., Giurgiu, I., Bogojeska, J., et al.: Predicting disk replacement towards reliable data centers. In: the 22nd ACM SIGKDD International Conference. ACM (2016)
3. Vishwanath, K.V., Nagappan, N.: Characterizing cloud computing hardware reliability. In: Proceedings of the 1st ACM Symposium on Cloud Computing, pp. 193–204 (2010)
4. Shen, J., Wan, J., Lim, S.J., et al.: Random-forest-based failure prediction for hard disk drives. Int. J. Distrib. Sensor Netw. **14**(11) (2018)
5. Jiang, T., Zeng, J., Zhou, K., et al.: Lifelong disk failure prediction via GAN-based anomaly detection. In: 2019 IEEE 37th International Conference on Computer Design (ICCD), pp. 199–207. IEEE (2019)
6. Kamarthi, S., Zeid, A., Bagul, Y.: Assessment of current health of hard disk drives. In: Proceedings of the Fifth Annual IEEE International Conference on Automation Science and Engineering. IEEE (2009)
7. Murray, J.F., Hughes, G.F., Kreutz-Delgado, K.: Machine learning methods for predicting failures in hard drives: a multiple-instance application. J. Mach. Learn. Res. **6**(1), 783–816 (2005)
8. Xiao, J., Xiong, Z., Wu, S., Yi Y., Jin, H., Hu, K.: Disk failure prediction in data centers via online learning. In: ICPP 2018: Proceedings of the 47th International Conference on Parallel Processing, pp. 1–10 (2018)
9. Li, J., Ji, X., Jia, Y., et al.: Hard drive failure prediction using classification and regression trees. In: 2014 44th Annual IEEE/IFIP International Conference on Dependable Systems and Networks (DSN), pp. 383–394. IEEE (2014)
10. Yang, W., Hu, D., Liu, Y., et al.: Hard drive failure prediction using big data. In: IEEE Symposium on Reliable Distributed Systems Workshop. IEEE (2015)

11. Xu, Y., Sui, K., Yao, R., et al.: Improving service availability of cloud systems by predicting disk error. In: 2018 USENIX Annual Technical Conference. USENIX (2018)

12. Eduardo, P., Wolfdietrich, W., Luiz, A.B.: Failure trends in a large disk drive population. In: USENIX Conference on File and Storage Technologies San Jose. USENIX (2007)

13. Gunawi, H.S., Hao, M., Suminto, R.O., et al.: Why does the cloud stop computing?: lessons from hundreds of service outages. In: Seventh ACM Symposium. ACM (2016)

14. Han, S., Wu, J., Xu, E., et al.: Robust data preprocessing for machine-learning-based disk failure prediction in cloud production environments. arXiv preprint arXiv:1912.09722 (2019)

15. Stoer, J., Bulirsch, R.: Introduction to Numerical Analysis, vol. 12, 3rd edn. Springer, New York (2002). https://doi.org/10.1007/978-0-387-21738-3

16. Zhu, B., Wang, G., Liu, X., et al.: Proactive drive failure prediction for large scale storage systems. In: 2013 IEEE 29th Symposium on Mass Storage Systems and Technologies (MSST), pp. 1–5. IEEE (2013)

17. He, H., Ma, Y.: Imbalanced Learning. Foundations, Algorithms, and Applications. Wiley, Hoboken (2013)

18. Hughes, G.F., Murray, J.F., Kreutz-Delgado, K., Elkan, C.: Improved disk-drive failure warnings. IEEE Trans. Reliab. **51**(3), 350–357 (2002)

19. Murray J.F., Hughes G.F., Kreutz-Delgado K.: Hard drive failure prediction using non-parametric statistical methods. In: ICANN/ICONIP 2003 Proceedings of Joint International Conference on Artificial Neural Networks and Neural Information Processing. Springer, Heidelberg (2003)

20. Hamerly, G., Elkan, C.: Bayesian approaches to failure prediction for disk drives. In: ICML 2001 Proceedings of the Eighteenth International Conference on Machine Learning, pp. 202–209 (2001)

21. Murray, J.F., Hughes, G.F., Kreutz-Delgado, K.L.: Machine learning methods for predicting failures in hard drives: a multiple-instance application. J. Mach. Learn. Res. **6**, 783–816 (2005)

22. Ma, A., Douglis, F., Lu, G., et al.: RAIDShield: characterizing, monitoring, and proactively protecting against disk failures. In: Proceedings of the 13th USENIX Conference on File and Storage Technologies, pp. 241–256. SENIX Association (2015)

23. Qian, J., Skelton, S., Moore, J., et al.: P3: priority based proactive prediction for soon-to-fail disks. In: 2015 IEEE International Conference on Networking, Architecture and Storage (NAS), pp. 81–86. IEEE (2015)

24. Wang, Y., Miao, Q., Ma, E.W., et al.: Online anomaly detection for hard disk drives based on Mahalanobis distance. IEEE Trans. Reliab. **62**(1), 136–145 (2013)

25. Wang, Y., Ma, E.W., Chow, T.W., et al.: A two-step parametric method for failure prediction in hard disk drives. IEEE Trans. Industr. Inf. **10**(1), 419–430 (2014)

26. Queiroz, L.P., Rodrigues, F.C.M., Gomes, J.P.P., et al.: A fault detection method for hard disk drives based on mixture of Gaussians and nonparametric statistics. IEEE Trans. Industr. Inf. **13**(2), 542–550 (2017)

27. Xu, C., Wang, G., Liu, X.G., et al.: Health status assessment and failure prediction for hard drives with recurrent neural networks. IEEE Trans. Comput. **13**(9), 1–8 (2015)

28. Lu, S., Luo, B., Patel, T., et al.: Making disk failure predictions SMARTer!. In: 18th USENIX Conference on File and Storage Technologies, pp. 151–167. USENIX (2020)

29. Goodfellow, I., Pouget-Abadie, J., Mirza, M., et al.: Generative adversarial nets. In: Advances in Neural Information Processing Systems, pp. 2672–2680 (2014)
30. Prokhorenkova, L., Gusev, G., Vorobev, A., et al.: CatBoost: unbiased boosting with categorical features. In: Advances in Neural Information Processing Systems, pp. 6638–6648 (2018)

PAKDD2020 Alibaba AI Ops Competition: An SPE-LightGBM Approach

Yuanpeng Li[1] and Yaoran Sun[1,2]([✉])

[1] Zhejiang University, Hangzhou 310058, China
{yuanpengli,sunoi}@zju.edu.cn
[2] Hangzhou Zhuxing Information Technology Co., Ltd., Hangzhou 310058, China

Abstract. This paper describes our submission to the PAKDD2020 Alibaba AI Ops Competition. We regard the hard driver disk failure prediction problem as a binary classification problem. Our approach is based on self-paced ensemble (SPE) [9] and a light gradient boosting machine (LightGBM) [8]. With three types of feature (raw feature, window-based feature and combined raw feature) and our proposed training sample selection strategy, our approach achieved rank 14 in the final standings with F-score (defined in [1]) of 0.37. The code for our approach can be found in https://github.com/fengyang95/Alibaba_AI_Ops_Competition_Rank14.

Keywords: Feature engineering · Random under sampling · Self-paced ensemble · LightGBM

1 Introduction

The goal of the PAKDD2020 Alibaba AI Ops Competition was to determine whether each hard disk drive (HDD) will fail within the next 30 days. The datasets [2] consist of two parts, fault label data and a period of time of daily disk status monitoring data (Self-Monitoring, Analysis, and Reporting Technology; often written as SMART).

We transformed this failure prediction problem into a traditional binary classification task, and labeled the logs of the disk that will have a fault record within 30 days as 1, and the remaining logs as 0. There are several major difficulties in this competition. The first is the extreme imbalance of logs. In the dataset from July 2017 to August 2018, there are about 56,000,000 negative logs, but only 38551 positive logs and the ratio of positive and negative logs reached 1:1500. Then there is the noise problem. The dataset comes from the actual data of the industry, there is a lot of noise and it is easy to cause overfitting. Finally, there is the problem of the amount of data. A total of 56,000,000 logs of about 42 gigabyte (GB) of data have also caused certain difficulties in data processing.

Supported by Alibaba.

Our approach contains 3 main steps:

- Feature engineering.
- Applying random down sampling on negative logs.
- Using self-paced ensemble (SPE) with LightGBM to train a binary classifier.

The remainder of this article is organized as follows. Section 2 will give a brief description of the background and related work. The details of our approach and experimental results will be described in Sect. 3. Section 4 summarizes the entire article with some discussions.

2   Related Work

In 1995, the drive industry adopted SMART: a standardized specification for HDDs failure warnings [7]. SMART is a built-in HDD function that works by calculating attribute values and is used to evaluate the performance of HDDs [12,15]. It reflects the health status of the disk by monitoring and reading the operating data (such as temperature, raw reading error rate, start/stop count, power on time count, etc.) of hard disk's heads, platters, and circuits.

In recently, many machine learning methods have been applied to SMART-based failure prediction task, including BPNN [17], priority-based proactive prediction (P3) [13], RNN-based [16], Random-forest-based [14] etc. There are also many statistical approaches, for example, multivariate rank-sum test and OR-ed single variate test [17], Bayesian approaches [5], RAIDShield [10] etc. In general, the method based on machine learning can achieve better results, and our SPE-LightGBM approach is also one of them.

3   Method

3.1   Feature Engineering

We mainly use single log data to make predictions, which include three aspects of features: raw features, window features and combined features.

Raw Features. After simple data analysis and removing some features with many missing values, we used the following original features as shown in Table 1.

Window Features

- range_smart_1_normalized, std_smart_1_normalized
- range_smart_5raw, std_smart_5raw
- range_smart_7_normalized, std_smart_7_normalized
- range_smart_190raw, std_smart_190raw
- range_smart_191raw, std_smart_191raw
- range_smart_193raw, std_smart_193raw

Table 1. Raw features

Feature	Description
smart_1_normalized	Raw read error rate
smart_3_normalized	Spin up time
smart_4raw	Start/Stop count
smart_5raw, smart_5_normalized	Reallocated sector count/Retired block count
smart_7raw, smart_7_normalized	Seek error rate
smart_9raw, smart_9_normalized	Power-on time count (POH)
smart_12raw	Power cycle count
smart_187raw, smart_187_normalized	Reported uncorrectable erros
smart_188raw, smart_188_normalized	Command timeout
smart_190raw, smart_190_normalized	Airflow temperature
smart_191raw, smart_191_normalzied	G-sense error rate
smart_192raw	Power-off retract count
smart_193raw	Load/Unload cycle count
smart_194raw, smart_194_normalized	Temperature
smart_195raw, smart_195_normalized	
smart_197raw, smart_197_normalized	Current pending sector count
smart_198raw, smart_198_normalized	Total count of read sectors
smart_199raw	Total count of write sectors

- range_smart_194raw, std_smart_194raw
- range_smart_195raw, std_smart_195raw
- range_smart_195_normalized, std_smart_195_normalized
- range_smart_199raw, std_smart_199raw

For some smart attributes, when the value changes suddenly, it means that the disk has meet some extreme conditions, which may have an impact on the life of the disk. We have added some statistical window features on the basic of the original features, including the numerical variation range and standard deviation within a 7-days time window.

Where the features with 'range' prefix represent the difference between the maximum and minimum values in the time window, and the 'std' prefix represent the standard deviation of the data in the time window.

Combined Features. In addition to the raw features and window features, the combination of raw features are also considered. We have selected 6 important raw features according to the feature importance rank using random forests [3]. Then the two-by-two combinations of these 6 features (smart_4raw, smart_5raw, smart_187raw, smart_191raw, smart_197raw, smart_198raw) are used to generate 15 new combined features.

Algorithm 1. Generate Combined Features

Input: The set of selected raw features F_n, the number of selected raw features n.
Output: The set of combined features C_n
1: $k = 1$;
2: **for** $i = 1; i < n; i + +$ **do**
3:     **for** $j = i + 1; j < n; j + +$ **do**
4:         $C_n[k] = log(F_n[i] + 1) + log(F_n[j] + 1)$;
5:         $k = k + 1$;
6:     **end for**
7: **end for**

Table 2. Comparison of different features

Features types	Number of features	AUC	AUCPR
Raw features	29	**0.739**	0.041
Window features	20	0.673	0.049
Combine features	15	0.671	0.037
Raw + window features	49	0.725	0.065
Raw + combined features	44	0.735	0.044
Window + combined features	35	0.711	0.066
Raw + window + combined features	64	0.727	**0.068**

Experiments. The offline results of several features are compared to evaluate the effectiveness of feature engineering. The model used in the experiment is SPE with LightGBM as its base estimator, which will be introduced in detail in Sect. 3.3. Training samples are selected according to sample1 described in the following Sect. 3.2. The area under receiver operating characteristic curve (AUC) and area under precision-recall curve (AUCPR) were used to evaluation. From Table 2, it can be seen that our three-group feature (including raw features, window features and combined features) approach achieved the best results (with AUCPR 0.068). If only the raw features are used, the AUCPR is 0.041, indicating the effectiveness of the feature engineering of our approach.

3.2   Training Sample Selection

According to our experiments, the selection of training samples has a great influence on the performance of the model. Due to the limitations of machine memory and efficiency considerations, we did not use all samples for training, but did some preprocessing. We have tried several sample selection methods, and finally decided to use the following strategy to construct the training set: for negative samples, we randomly select 2 logs from dozens of logs per month per disk, and then compose them together with all positive samples to form a training set.

We conducted offline comparison experiments to verify the effectiveness of this method of selecting samples. We have used the logs data from August 1, 2017 to August 1, 2018 for offline experiments, among them the logs data after July 1, 2018 was used as the validation set. Four methods for selecting training samples were compared and the results are shown in the Table 3.

Table 3. Sampling strategy

Sampling	Description	Number of training samples
Sample1	Randomly select 2 logs per month per disk for negative samples with all positive samples from August 1, 2017 to July 1, 2018	P:27907 N:2829070
Sample2	Only use negative samples from April 1 to June 1, 2018 with all positive samples from August 1, 2017 to July 1, 2018	P:27907 N:9106168
Sample3	Only use the samples corresponding to the disk with fault tag	P:27907 N:120752
Sample4	All positive and negative samples from April 1 to June 1	P:6505 N:9106168

It can be seen from Fig. 1 that using sample1 can lead to the best results which achieves area under ROC curve 0.733 and area under precision-recall curve (AUCPR) 0.077.

3.3 Model

We use a computationally efficient model called self-paced ensemble (SPE) [9] with LightGBM [8] as its base estimator.

SPE is a novel framework for imbalance classification that aims to generate a strong ensemble by self-paced harmonizing data hardness via down-sampling. It considers the distribution of classification hardness over the dataset and iteratively selects the most informative majority data samples according to the hardness distribution. A self-paced procedure is used to control this down-sampling strategy which enables the framework focuses on the harder data samples. The SPE framework has been well applied in many unbalanced data scenarios. Light-GBM [8] is a new gradient boosting decision tree (GBDT) implementation with gradient-based one-side sampling (GOSS) and exclusive feature bundling (EFB) strategy.

In our approach, training samples are selected according to sample1 described in Sect. 3.2. We compared several baseline methods which are listed in Table 4. It can be seen from Fig.2 that the SPE-LightGBM method is the best (with AUCPR 0.068). Actually, the SPE-LightGBM here uses the same scheme as the sample1 in Sect. 3.2, but due to the random seed, there is a slight difference between the two results.

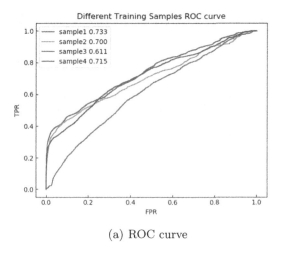

(a) ROC curve

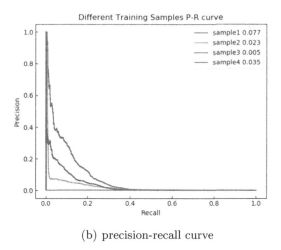

(b) precision-recall curve

Fig. 1. ROC curve and precision-recall curve using different training samples.

Table 4. Baselines

Model	Description	Main parameters setup
Logistic Regression (LR) [11]	A classical linear classification model	L2 regularization; max_iter = 100
AdaBoost [4]	An ensemble method that constructs a classifier in an iterative fashion	max_iter set to 50 with decision tree as base estimator
Random Forest (RF) [3]	A bagging model the core idea of which is to generate multiple small decision trees from random subsets of the data	max_iter = 100

(contniued)

Table 4. *(contniued)*

LightGBM	–	num_leaves = 127, learning rate set to 0.001
balanced-LightGBM	LightGBM with weights inversely adjustment proportional to class frequencies in the input data	num_leaves = 127, learning rate set to 0.001, is_unbalance set to true flag
SPE-LightGBM	SPE with LightGBM as base estimator	n_estimators = 20
SPE-GBDTLR	SPE with GBDT + LR [6] as base estimator	n_estimators = 20

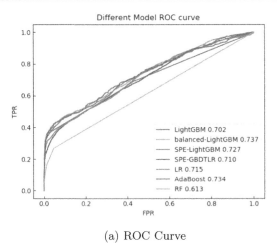

(a) ROC Curve

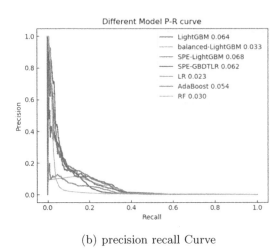

(b) precision recall Curve

Fig. 2. ROC curve and precision-recall curve using different models.

4 Conclusions

We use raw features, window features and combined features, then undersample the negative samples, and finally use SPE-LightGBM to train a classifier. After choosing an appropriate classification threshold, our model obtained an F-score of 0.37 (with precision 0.424 and recall 0.328) on the online test set (disk logs for September 2018).

We have constructed such a complete pipline that predicts disk failure, which may bring some inspiration to AI Ops. Unfortunately, our approach does not handle noise well. Perhaps adding more fine-grained labels, for example, turning this problem into a regression task may achieve better prediction results.

Acknowledgements. Thanks to Alibaba and PAKDD for hosting, creating and supporting this competition.

References

1. https://tianchi.aliyun.com/competition/entrance/231775/information?lang=en-us
2. https://github.com/alibaba-edu/dcbrain
3. Breiman, L.: Random forests. Mach. Learn. **45**(1), 5–32 (2001)
4. Freund, Y., Schapire, R., Abe, N.: A short introduction to boosting. J. Jpn. Soc. Artif. Intell. **14**(771–780), 1612 (1999)
5. Hamerly, G., Elkan, C., et al.: Bayesian approaches to failure prediction for disk drives. In: ICML, vol. 1, pp. 202–209 (2001)
6. He, X., et al.: Practical lessons from predicting clicks on ads at Facebook. In: Proceedings of the Eighth International Workshop on Data Mining for Online Advertising, pp. 1–9 (2014)
7. Hughes, G.F., Murray, J.F., Kreutz-Delgado, K., Elkan, C.: Improved disk-drive failure warnings. IEEE Trans. Reliab. **51**(3), 350–357 (2002)
8. Ke, G., et al.: LightGBM: a highly efficient gradient boosting decision tree. In: Advances in Neural Information Processing Systems, pp. 3146–3154 (2017)
9. Liu, Z., Cao, W., Gao, Z., Bian, J., Chen, H., Chang, Y., Liu, T.Y.: Self-paced ensemble for highly imbalanced massive data classification. In: 2020 IEEE 36th International Conference on Data Engineering (ICDE). IEEE (2020)
10. Ma, A., Traylor, R., Douglis, F., Chamness, M., Lu, G., Sawyer, D., Chandra, S., Hsu, W.: RAIDShield: characterizing, monitoring, and proactively protecting against disk failures. ACM Trans. Storage (TOS) **11**(4), 1–28 (2015)
11. Menard, S.: Applied Logistic Regression Analysis, vol. 106. Sage, Thousand Oaks (2002)
12. Murray, J.F., Hughes, G.F., Kreutz-Delgado, K.: Machine learning methods for predicting failures in hard drives: a multiple-instance application. J. Mach. Learn. Res. **6**(May), 783–816 (2005)
13. Qian, J., Skelton, S., Moore, J., Jiang, H.: P3: priority based proactive prediction for soon-to-fail disks. In: 2015 IEEE International Conference on Networking, Architecture and Storage (NAS), pp. 81–86. IEEE (2015)
14. Shen, J., Wan, J., Lim, S.J., Yu, L.: Random-forest-based failure prediction for hard disk drives. Int. J. Distrib. Sens. Netw. **14**(11), 1550147718806480 (2018)

15. Wang, Y., Miao, Q., Pecht, M.: Health monitoring of hard disk drive based on Mahalanobis distance. In: 2011 Prognostics and System Health Management Conference, pp. 1–8. IEEE (2011)
16. Xu, C., Wang, G., Liu, X., Guo, D., Liu, T.Y.: Health status assessment and failure prediction for hard drives with recurrent neural networks. IEEE Trans. Comput. **65**(11), 3502–3508 (2016)
17. Zhu, B., Wang, G., Liu, X., Hu, D., Lin, S., Ma, J.: Proactive drive failure prediction for large scale storage systems. In: 2013 IEEE 29th Symposium on Mass Storage Systems and Technologies (MSST), pp. 1–5. IEEE (2013)

Noise Feature Selection Method in PAKDD 2020 Alibaba AI Ops Competition: Large-Scale Disk Failure Prediction

Yifan Peng[1](✉), Junfeng Xu[2], and Nan Zhao[3]

[1] School of Computer Science and Engineering,
University of Electronic Science and Technology of China, Chengdu, China
yifannir@gmail.com
[2] Agricultural Bank of China, Xiamen Branch, Xiamen, China
donaldxu@gmail.com
[3] Agricultural Bank of China, Hangzhou Branch, Hangzhou, China
mr.zhaonan@outlook.com

Abstract. This paper describes our method to the PAKDD2020 Alibaba AI Ops Competition: Large-Scale Disk Failure Prediction. Our approach is based on Gradient Boosting Machine and deep feature engineering, and most important is the research on feature selection method. In this competition, we proposed a new feature selection method based on the existing Null Importance method. We named Noise Feature Selection short for NFS. To fitted to noise, NFS using target permutation tests actual significance against the whole distribution of feature importance. The effectiveness of NFS method has been proved in experiments. While in the competition task, we got 0.2509 score in Qualification, 0.1946 score in Semi-Finals.

Keywords: Disk Failure Prediction · Feature selection · Feature engineering

1 Introduction

In the large-scale data centers, the number of hard disk drive (HDD) and solid-state drive (SSD) has reached millions. According to statistics, disk failures account for the largest proportion of all failures. The frequent occurrence of disk failures will affect the stability and reliability of the server and even the entire IT infrastructure, which have a negative impact on business Service-Level Agreement. Thus, prediction of disk failures has been an important topic for IT or big data company.

Hard drive manufacturers have been developing self-monitoring technology in their products since 1994, in an effort to predict failures early enough to allow users to do something worth of their data. This Self-Monitoring and Reporting

C. He et al. (Eds.): AI Ops 2020, CCIS 1261, pp. 109–118, 2020.
https://doi.org/10.1007/978-981-15-7749-9_11

Technology (SMART) system uses attributes collected during normal operation to set a failure prediction flag. And this PAKDD2020 Alibaba AI Ops Competition: Large-Scale Disk Failure Prediction is based on this type of SMART datasets [1].

In terms of feature generation of structured data, On the one hand, it is mostly based on the feature discovery of the business to which the data belongs. On the other hand, new features can be generated by statistical methods or timing methods, etc. After generating many features, feature selection becomes an extremely important task. If feature selection is not performed, the model will overfit and learn some useless correlations of noise features [3]. If the feature selection is not good, It may cause under fitting, and the algorithm cannot learned well.

Several methods have been proposed in the past decades to improve the accuracy of HDD failure prediction. Most algorithms apply statistical approaches, machine learning, and deep learning technologies, including rank-sum test, naive Bayesian classifiers [5], Hidden Markov Model and Hidden Semi-Markov Model approaches [13], classification tree [9], and neural networks [12] etc.

However, the focus of these prediction methods is on the optimization of the model, and very few focus on feature selection and feature engineering. In this paper, we propose a new feature selection method based on the defects of existing method. And the experiments on the hard disk detection task show that our feature selection method has an improvement on performance.

In this paper, we first introduce the mathematical description of disk failure prediction problem, then briefly introduce the approach of our solution. Then noise feature selection method are proposed in the article, and finally verify the validity of our model and feature selection method through extensive experiments.

2   Problem Statement

The Disk Failure Prediction problem gives a dataset which is a period of time of daily disk status monitoring data (SMART data) and fault label data. The task is that we should daily determine whether each disk will fail within the next 30 days. In this paper, what is different from this competition is that we use full-scale data, without any truncation processing or sampling processing, so that it can better reflect the real situation. We defined this problem by model approach method. It can be defined as a classification problem.

$$Y = f(X, \theta)$$

where θ is the parameter of our classification function, X is SMART data record and Y is our prediction.

2.1   Data Description

The dataset has two files: daily SMART data and tag file, which ranges from 2017-07-31 to 2018-08-31[1]. The daily SMART data of disks that has 514 columns.

---

[1] https://github.com/alibaba-edu/dcbrain/tree/master/diskdata

Including disk serial number code, disk manufacturer code, disk model code, normalized SMART data of SMART ID = n, raw SMART data of SMART ID = n, sampling time, and Tag file also contains fault time of disk, IDs of fault subtype. As Table 1, 2 shows,

Table 1. SMART data.

serial_number	manufacturer	model	smart_n_normalized	smart_nraw	dt
disk_100056	A	1	81.0 (n = 1)	151788225.0 (n = 1)	2017-08-29
disk_143883	A	1	75 (n = 1)	34831589.0 (n = 1)	2017-08-29
...	...	...	...	...	...

Table 2. Fault data.

manufacturer	model	serial_number	fault_time	tag
A	1	disk_100102	2017-09-29	0
A	1	disk_119991	2017-09-30	1
...	...	...	...	...

2.2 Evaluation Metrics

According to our purpose of failure prediction that predicting whether each disk will fail or not within next a few days, we redefine the precision, recall and F-score metrics. The complete definition of metrics is as follows:

Precision for 30-day observation window

$$Precision = \frac{n_{tpp}}{n_{pp}}$$

Recall for 30-day observation window

$$Recall = \frac{n_{tpr}}{n_{pr}}$$

F-score for 30-day observation window

$$F_{score} = 2 \times \frac{Precision \times Recall}{Precision + Recall}$$

where n_{pp} is the number of disks that are predicted to be faulty in the following 30 days in the observation window. n_{tpp} is the number of all the disks those truly fail among 30 days after the first predicting day in the observation window. n_{pr} is the number of all the disk failures occurring in the k-day observation window. n_{tpr} is the number of truly faulty disks that are successfully predicted no more than 30 days in advance.

3   Disk Failure Prediction Approach

We divide the entire process of hard disk failure prediction into four parts, data preprocessing, feature engineering, feature selection, and model training. As we propose a new feature selection method in the feature selection, which will be described it separately in Sect. 4.

3.1   Data Process

In the data preprocessing part, we will perform preliminary feature selection on the data.

We delete fields with NaN greater than 90%, Select Raw type features, At the same time, according to the actual field meaning and considering the research results of other scholars. BackBlaze analysis found that hard disk failure has a great relationship with SMART's 5, 187, 188, 197, 198 attributes, El-Shimi analysis report pointed out that in the random forest model 9, 193, 194, 241 and 242 feature significant weight [4]. This prior knowledge and some of selection rules such as deleting features with Nan greater than 90% are used to filter original features. The preliminary feature selection process is shown in Fig. 1.

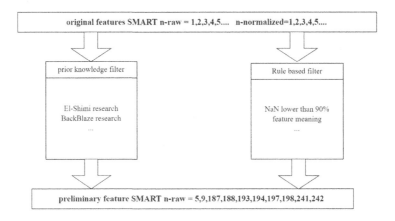

Fig. 1. Preliminary feature selection process

In summary, we use these preliminary features and some other features for data preprocessing. The following list gives some examples,

– SMART 5,9,187,188,193,194,197,198,241,242 raw.

3.2   Feature Engineering

Through the features of the business and the SMART data features studied by other scholars, we have deleted some useless features after data preprocessing.

And the other features we called it basic features. In this part, it is mainly divided into statistical features, cross features, and time series sliding window features.

Cross Features. In terms of Cross feature construction. We cross-combine basic features, so we can get higher-dimensional features of this data. The specific description is as follows,

$$Feature_{cross1(x,y)} = log(1 + Feature_x \times Feature_y) \tag{1}$$

$$Feature_{cross2(x,y)} = \frac{Feature_x}{Feature_y} \tag{2}$$

where $Feature_{cross1(x,y)}$, $Feature_{cross2(x,y)}$ is Feature x cross with Feature y with two ways.

Time Series Features. We use sliding windows to count time series feature of data. The sliding window is time dependent and the size is 10 days, user 'key' column to group data, and we make statistics of the time series feature of the first 10 days, like sum, min, means, median, kurt, skew operation. These indicators can express the distribution of slide window data.

3.3 Training Model

During many experiments we found that the tree method is suitable for this task. In this paper, we use LightGBM as our main algorithm. The detailed description of this method can be found in [7]. LightGBM is a gradient boosting tree framework, which is highly efficient and scalable. It can support many different algorithms including GBDT, GBRT, GBM, and MART. We use GBDT as its algorithms. Also, LightGBM is evidenced to be several times faster than existing implementations of gradient boosting trees, due to its fully greedy tree-growth method and histogram-based memory and computation optimization [2]. We use LightGBM as our detail model.

4   Noise Feature Selection Method

The process of feature selection includes not only deciding which attributes to use in the classifier, but also it is an important method to avoid overfitting. This part is feature selection method. Based on null importance feature selection, We made some improvements on null importance feature selection, and proposed a new feature selection method which called it Noise Feature Selection Method (NFS). In this paper, We conduct comparison experiments through variance feature selection, mutual information feature selection and null importance feature selection.

4.1 Variance Selection

Variance selection method (VS) is a simple baseline approach to feature selection. It removes all features whose variance does not meet some threshold. It can be write as,

$$F_{varselect} = \{var(f) > t_{var} | f \in F_{all}\} \qquad (3)$$

Where $F_{varselect}$ is the features we selected by variance selection. $var(f)$ is the variance of feature f. t_{var} is the threshold.

4.2 Mutual Information Selection

Mutual information selection method (MI) calculate by estimate mutual information for target variable and features [8]. And removes all features whose mutual information does not meet some threshold with the target. It can be write as,

$$F_{mulselect} = \{mul(f, target) > t_{mul} | f \in F_{all}\} \qquad (4)$$

Where $F_{mulselect}$ is the features we selected by mutual information selection. $mul(f, target)$ is the mutual information between feature f and target. t_{mul} is the threshold.

4.3 Null Importance Feature Selection

Null importance feature selection (NI) process using target permutation tests actual importance significance against the distribution of feature importance when fitted to noise (shuffled target). and it uses the log actual feature importance divided by the some percentiles (usually 75 percentile) of null distribution.

$$F_{niselect} = \{actual(f) > p(shuffle(f)) | f \in F_{all}\} \qquad (5)$$

Where $F_{niselect}$ is the features we selected by null importance feature selection. $actual(f)$ is the actual feature importance score which calculated by a model (We use random forest here [10]). $Shuffle(f)$ is feature importance when fitted to noise. p function is the choice strategy. We use 75 percentile in this paper.

4.4 Noise Feature Selection

Noise Feature Selection Method (NFS) is based on null importance feature selection, But the difference is that the NFS is the score under the real distribution score, which is more than the some percentile of the overall random score distribution. Instead of comparing the true distribution score with the random distribution score for each feature.

For we think that in the large number of feature group spaces generated after feature engineering, some features are invalid by themselves, regardless of whether they have undergone random distribution training. These noise characteristics should be excluded. Therefore, the viewpoint of NI should be combined with the viewpoint of eliminating features with a small contribution degree, that is, to delete features that do not meet the threshold value of the overall contribution score as Eq. 6 shows,

$$F_{nfsselect} = \{actual(f) > t_{allshuffle(f)} | f \in F_{all}\} \tag{6}$$

Where $F_{nfsselect}$ is the features we selected by noise feature selection. $actual(f)$ is the actual feature importance score as Eq. 5. $allshuffle(f)$ is feature importance distribution when fitted to noise. $t_{allshuffle(f)}$ is the threshold of our choice strategy.

5 Experiments

In the experimental part, we set up two sets of experiments. The first group of experiments is a feature group comparison experiment, the purpose is to get the performance of each feature group on the task. The second group of experiments is a feature selection comparison experiment. We use different feature selection methods on different feature groups to compare our proposed feature selection methods.

5.1 Experiments Setting

We use LightGBM after tuning as a training model, And use different feature groups generated by feature engineering, basic feature groups, cross feature groups, time-series feature groups for the feature space input of the model.

In terms of dataset, we use data from 2018-05 to 2018-06 SMART dataset for training and 2018-07 data for testing and verification.

5.2 Feature Comparison Experiment

Table 3. Feature performance comparison without feature selection.

Feature group	Recall	Precision	F score
Basic feature	0.1026	0.1338	0.1161
Cross	0.1124	0.1434	0.1260
Time series	**0.1735**	0.2037	0.1874
Cross and time series	0.1539	**0.2239**	**0.1824**

As Table 3 can be show, from a macroscopic observation, On these sub datasets, time series feature group shows the best effect. And the basic feature without any transform shows the worst effect. This shows in disk failure prediction task, time series features are very effective.

5.3   Comparison Experiment of Feature Selection

Table 4. Feature selection performance comparison.

Feature group	VS			MI		
	Recall	Precision	F score	Recall	Precision	F score
Basic	0.1231	0.1592	0.1388	0.1331	0.1594	0.1451
Cross	0.1144	0.1572	0.1324	0.1241	0.1777	0.1461
Time series	0.1832	0.2131	0.1970	0.1861	0.2042	0.1947
Cross and time series	0.1742	0.2240	0.1960	0.1701	0.2253	0.1938
	NI			NFS		
	Recall	Precision	F score	Recall	Precision	F score
Basic feature	0.1492	0.2131	**0.1755**	0.1440	0.2010	0.1678
Cross	0.1504	0.2107	0.1755	0.1924	0.2213	**0.2058**
Time series	0.1929	0.2440	0.2155	0.2021	0.2513	**0.2240**
Cross and time series	0.2146	0.2491	0.2306	0.2303	0.2657	**0.2467**

In order to compare the effectiveness of different feature selection method, we verified it through a comparison experiment of changing the feature selection method on the different feature groups. It can be seen from Table 4 that our NFS method shows the best score at most cases, whether in basic features or some other feature groups. As can be seen from the Table 4, NFS has not achieved the best results in the basic feature group. As the number of features in the basic feature group is small, NFS has a good selection effect among a large number of features and features containing many noise. Indicating that the NFS method is still active when the feature groups change.

Comparing the F1 score among different feature selection methods, as Fig. 2 shows, Null importance and NFS method has brought a significant improvement. When the feature is basic group, the NFS method may not be as good as other methods, but when the number of features is large like cross feature, NFS is effective.

This part show our selection method have a high performance on different feature groups. Saying that NFS method based on null importance can be useful.

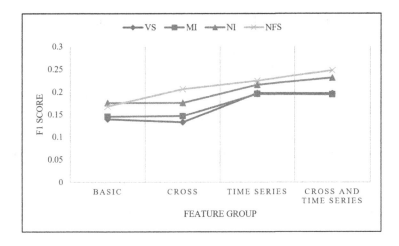

Fig. 2. F1 score between different feature selection method

6 Conclusion and Future Work

At this Large-Scale Disk Failure Prediction Competition. We proposed a new feature selection method based on the existing null importance method. And this method has proved effective in experiments. In disk failure prediction task, disk fault feature mining, there is still a lot of research to do. Because of time issues, there are many methods in the paper and some exploratory attempts that were not used in the competition. Therefore, in the following experiments, the experimental results on the full amount of data were better than the truncated data in the competition.

In the follow-up research, There are many directions to be tried. As shown in the following list,

- Applicability of model. We plan to use neural network series such as Long short-term memory neural network (LSTM) [6] and other time series models to increase the applicability of the model.
- Concept Drift. For the concept drift problem in the time dimension, continuous learning models such as state-of-the-art decision tree classification method (CVFDT) and Efficient CVFDT (E-CVFDT) [11] can be considered.
- Feature Mining. On feature mining, feature bucketing can be embed to improve the mining ability of SMART data.

References

1. PAKDD2020 Alibaba AI OPS competition: Large-scale disk failure prediction (2020). https://www.pakdd2020.org/competition_aiops.html
2. Essam Al Daoud: Comparison between XGBoost, LightGBM and CatBOOST using a home credit dataset. Int. J. Comput. Inform. Eng. **13**(1), 6–10 (2019)

3. Dash, M., Liu, H.: Feature selection for classification. Intell. Data Anal. **1**(3), 131–156 (1997)
4. El-Shimi, A.: Predicting storage failures (2017)
5. Hamerly, G., Elkan, C.: Bayesian approaches to failure prediction for disk drives, pp. 202–209 (2001)
6. Hochreiter, S., Schmidhuber, J.: Long short-term memory. Neural Comput. **9**(8), 1735–1780 (1997)
7. Ke, G., et al.: LightGBM: a highly efficient gradient boosting decision tree. In: Advances in Neural Information Processing Systems, pp. 3146–3154 (2017)
8. Kraskov, A., Stögbauer, H., Grassberger, P.: Estimating mutual information. Phys. Rev. E **69**(6), 066138 (2004)
9. Li, J., Stones, R.J., Wang, G., Liu, X., Li, Z., Xu, M.: Hard drive failure prediction using decision trees. Reliab. Eng. Syst. Saf. **164**, 55–65 (2017)
10. Liaw, A., Wiener, M., et al.: Classification and regression by randomforest. R News **2**(3), 18–22 (2002)
11. Liu, G., Cheng, H.-R., Qin, Z.-G., Liu, Q., Liu, C.-X.: E-CVFDT: an improving CVFDT method for concept drift data stream. In 2013 International Conference on Communications, Circuits and Systems (ICCCAS), vol. 1, pp. 315–318. IEEE (2013)
12. Chang, X., Wang, G., Liu, X., Guo, D., Liu, T.-Y.: Health status assessment and failure prediction for hard drives with recurrent neural networks. IEEE Trans. Comput. **65**(11), 3502–3508 (2016)
13. Zhao, Y., Liu, X., Gan, S., Zheng, W.: Predicting disk failures with HMM- and HSMM-based approaches. In: Perner, P. (ed.) ICDM 2010. LNCS (LNAI), vol. 6171, pp. 390–404. Springer, Heidelberg (2010). https://doi.org/10.1007/978-3-642-14400-4_30

Characterizing and Modeling for Proactive Disk Failure Prediction to Improve Reliability of Data Centers

Tuanjie Wang, Xinhui Liang, Quanquan Xie, Qiang Li[(✉)], Hui Li,
and Kai Zhang

State Key Laboratory of High-end Server and Storage Technology,
Beijing, China
{wangtuanjie,liangxinhui,xiequanquan,
li.qiangbj}@inspur.com

Abstract. In modern datacenter, hard disk drive has the highest failure rate. Current storage system has data protection feature to avoid data loss caused by disk failure. However, data reconstruction process always slows down or even suspends system services. If disk failures can be predicted accurately, data protection mechanism can be performed before disk failures really happen. Disk failure prediction dramatically improve the reliability and availability of storage system. This paper analyzes disk SMART data features in detail. According the analysis results, we design an effective feature extraction and preprocessing method. And we have optimized the XGBoost's hyperparameters. Finally, ensemble learning is applied to further improve the accuracy of prediction. The experimental results of Alibaba data set show that our system predict disk failures within 30 days. And the F-score achieves 39.98.

Keywords: XGBoost · Feature engineering · Hyperparameter tuning · Ensemble learning

1 Introduction

Large-scale data center usually has millions of hard disks. Disk failure will decrease the stability and reliability of the storage system. And it may even endanger the entire IT infrastructure, and affect the business SLA. If disk failures were predicted in advance, data can be backed up or migrated during the spare time. Disk failure prediction can greatly reduce data loss and effectively improve the reliability of the data center.

SMART (Self-Monitoring, Analysis and Reporting Technology) [1] is a monitoring data supplied by HDD, solid-state drives (SSDs), and eMMC drives. All modern HDD manufacturers support the SMART specification. Currently, it's common to predict disk failures using on SMART data and AI technology. A SMART datasets [2] was provided to contestants by PAKDD2020 Alibaba AI Ops Competition [3]. Our disk prediction model was verified on it.

© Springer Nature Singapore Pte Ltd. 2020
C. He et al. (Eds.): AI Ops 2020, CCIS 1261, pp. 119–129, 2020.
https://doi.org/10.1007/978-981-15-7749-9_12

There are a large amount of related work on predicting disk failures. For example, Hongzhang [4] proposed an active fault-tolerant technology based on the "acquisition-prediction-migration-feedback" mechanism. Sidi [5] proposed a method of combining disk IO, host IO and location information for fault prediction. Based on CNN and LSTM neural network algorithm, this method can extract features and train model automatically. Yong [6] proposed an online disk failures prediction method named CDEF. CDEF combine disk-level SMART signals and system-level signals. CDEF use a cost-aware rank model to select the top r disks that are most likely to have errors. Yanwen [7] proposed a disk failures prediction and interpretation method DFPE. By extracting relevant features, DFPE derives the prediction rules of the model. DFPE evaluates the importance of the features, then improves the interpretability of complex models. Ganguly [8] utilized SMART and hardware-level features such as node performance counter to predict disk failure. Ma [9] investigate the impact of disk failures on RAID storage systems and designed RAIDShield to predict RAID-level disk failures. Nicolas [10] uses SVM, RF and GBT to predict disk failures. And it reaches 67% recall. Tan [11] proposed an online anomaly prediction method to foresee impending system anomalies. They applied discrete-time Markov chains to model the evolving patterns of system features, then used tree augmented naive Bayesian to train anomaly classifier. Dean [12] proposed an Unsupervised Behavior Learning system, which leverages an unsupervised method self organizing map to predict performance anomalies. Wang [13] also proposed an unsupervised method to predict disk anomaly based on mahalanobis distance. Ceph [14] has disk fault prediction features. It needs to train SMART raw data for 12 days and use SVM [15] to predict disk failures.

However, due to the complexity of the actual production environment, noisy data, and other uncertainties, developing a disk failure prediction system that can be used in large-scale data centers is very challenging:

- The positive and negative samples are extremely imbalanced. The reason is system downtime caused by disk failure occurred infrequently. Actually, for small-scale or short-load disk storage systems, the number of failed disks is very small.
- The change of S.M.A.R.T. values is difficult to predict. According to our observation, S.M.A.R.T. values will change only when the disk is near failure, and sometimes change suddenly. In addition, when the disk is healthy, its S.M.A.R.T. value could be large and stable. Therefore, it cannot rely only on the absolute value of S.M.A.R.T.
- The generalization ability of prediction model is insufficient. There are a large number of disks of different models or even different manufacturers in the same data center. If the generalization ability of the prediction model is not strong, it is difficult to obtain high-performance prediction results.

The contributions of this article are as follows:

- Through data exploration, SMART range analysis, changepoint analysis and other methods, we found several SMART attributes that are strongly correlate to disk failure. We determine time series feature extraction method and sliding window size. We establish the principle of labeling positive and negative samples.

- In order to eliminate the differences in feature data distribution of different models, feature scaling is performed during data preprocessing. As a result, a unified model can simplify the deployment process, and improve the model generalization ability.
- During model training stage, firstly, we choose XGBoost [16] algorithm as base model, which is simple and efficient. Then we fine-tune model parameters. Finally, soft voting method is used to ensemble each sub-model, and further improving the prediction performance of the model.

The rest of this paper is organized as follows: In Sect. 2, we describes the proposed approach and details. The evaluation of our approach and experiment results are described in Sect. 3. Section 4 presents conclusion.

2 Solution

In this section, we present our disk failure prediction approach. Figure 1 shows the overview of the approach.

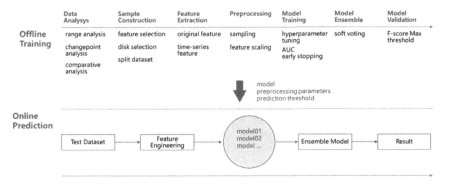

Fig. 1. Disk failure prediction overview.

Firstly, we analyze the internal distribution law of SMART data through data exploration, select representative healthy and faulty disks to construct positive and negative samples, identify fault-related SMART features and extract time series features. Feature scaling is performed during data preprocessing, and the impact of different ranges between different disk models and different SMART features can be eliminated. Secondly, based on the scaled dataset, we construct binary classification model and tune its hyperparameters. Finally, we integrate sub-models, verify integrated model using validation dataset, and take the threshold at the maximum F-score on the verification dataset as the optimal threshold.

We then import the trained model, preprocessing parameters and prediction threshold, and make online prediction on the test dataset and output the final prediction result.

2.1   Feature Selection

Through statistical analysis, we found that there are a large number of empty columns in the SMART dataset, and only 48 of the total 510 columns are non-empty. Then, the SMART probability density distribution and KL divergence were calculated for healthy and faulty disks, and SMART 5,187,192,193,197,198 and 199 were selected, which are related to disk failure and have big KL divergence. The KL divergence of all these selected features is positive infinity. As shown in Fig. 2, the KL divergence of SMART 198 is positive infinity, and the distribution of SMART 198 is mainly concentrated near zero for both healthy and faulty disks, and the main difference exists in the long tail on the right side. This part of data is useful for distinguishing between faulty disks and healthy disks. However, the distribution of SMART 194 has a high degree of coincidence, and the KL divergence is only 0.015, this means that it is difficult to distinguish between healthy and faulty disks through SMART 194.

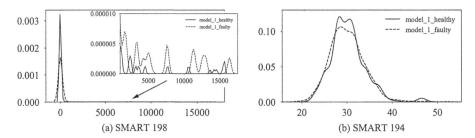

Fig. 2.   SMART 198 and 194 probability density distribution.

2.2   Feature Analysis

For the key smart features selected by the feature selection above, further analysis is made from the following three dimensions.

The first is range analysis. Statistics show that only around 5000 of the healthy disks exist non-zero value. Compared with all-zero disks, these disks contain more useful information. We should focus on these high-value healthy disk data when constructing model.

Secondly, changepoint analysis was performed on the SMART of faulty disks. It was found that even in the last 7 days of the faulty disks, the values of 50%–75% of these features such as SMART 5, SMART 187 are zero. And the faulty disk will not change significantly until the last 1–15 days of the life. As shown in Fig. 3, the SMART 5 of this disk did not change until the last 10 days, and did not increase significantly until the last 4 days, and SMART 187 did not change until the last 1 day. This phenomenon commonly occurs on faulty disks, that is, the closer to the end of life, the more likely sudden change will occur. Therefore, when constructing a positive sample, it is best to choose the last 0–7 days of the faulty disks, and the sliding window for extracting time series features is most suitably set between 3 and 7.

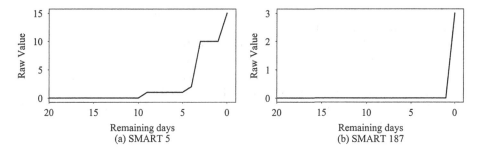

Fig. 3. SMART 5 (left) and 187 (right) trend graph.

Finally, by horizontal comparison and analysis of different disk models model1 and model2, it is found that the difference in the value range of each SMART feature between model 1 and model 2 is significant (As shown in Fig. 4). By scaling the SMART features of different models in the preprocessing stage, the SMART features of each model are firstly scaled to the same range, and then data is scaled again by standard scaler for training to eliminate smart distribution difference of each model. In this way, we obtain a unified model, optimize the prediction effect and improve model generalization ability successfully.

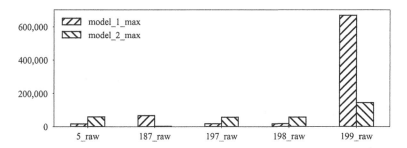

Fig. 4. SMART maximum comparison between model1 and model2.

2.3 Preprocessing

We use the dataset provided by Alibaba to complete our approach. The data from July 2017 to July 2018 is used for training, and the data of August 2018 is used for offline validation. Tianchi Alibaba uses data of September 2018 for online testing.

In the training dataset, there are a total of 184,305 disks, including 1,272 faulty disks and 183,033 healthy disks. Among all the disks, only 5,953 are not-all-empty. The judgment rule about not-all-empty is that the values of the main features (smart_5raw, smart_187raw, smart_197raw, smart_198raw, smart_199raw) are not all 0 or empty during the entire life cycle of the disk. For the training dataset, healthy and faulty disks are down-sampled at 10: 1, and around 5,000 not-all-empty healthy disks were added as supplements.

The missing values of the original data are filled with forward padding method to ensure the continuity of the time series.

To solve the problem of sample imbalance, we only select data samples for training from the last 7 days, and the last 30th, 40th, 50th, and 60th days of each disk. Then mark the last 7 days of the faulty disks as positive and other data samples as negative.

Time series feature extraction is performed on key SMART features on every day sampled. The sliding window is 3 days, 5 days, and 7 days. The extraction method is shown in the following Table 1.

Table 1. Time series feature extraction.

Number	Feature extraction method	Detail
1	Change time	The number of attribute changes within a period
2	Change rate	The slope of attribute values within a period
3	Std	The standard variance of attribute values within a period

Scale the SMART data of model2 to the range of the minimum and maximum of model1. Taking the feature Fn of model2 that is scaled to model1 as an example, first calculate the scaling factor, then scale the feature Fn of model2 to Fn_{scaled}.

$$scale = \frac{max(model1\_Fn) - min(model1\_Fn)}{max(model2\_Fn) - min(model2\_Fn)} \tag{1}$$

$$Fn_{scaled} = scale \times (Fn - min(model2\_Fn)) + min(model1\_Fn) \tag{2}$$

Finally, a standard method is used to scale the dataset.

2.4 Model Training

Our approach finally chose XGBoost [17] algorithm for model training, because the number of samples and the number of features in the data set are relatively small, and there is no need for very complicated models. At the same time, the hyperparameters of XGBoost are easy to adjust, and XGBoost is not easy to overfit. By comparing experimental results, it is found that the prediction results of XGBoost are better than Random Forest [18] and LSTM [18, 19].

We use 3-folder cross-validation for model training, and use AUC as the evaluation function. Compared with the PRC evaluation function, AUC is not sensitive to the rate of positive and negative samples. The AUC learning curve during training is shown in the Fig. 5. When the AUC is no longer improved, the optimal number of iterations of XGBoost can be determined.

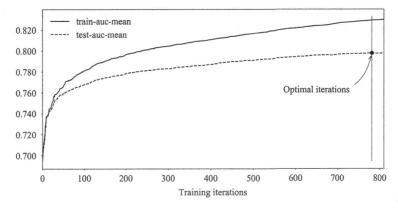

Fig. 5. AUC learning curve during training.

By using grid search to optimize XGBoost hyperparameters, such as max_depth, scale_pos_weight and so on, it was found that the prediction results were not improved significantly on validation dataset.

Finally, we use the validation dataset to obtain the prediction probability. As shown in the Fig. 6, the best prediction threshold is the classification threshold with maximum F-score.

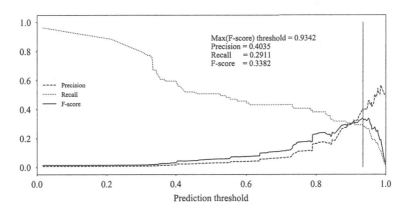

Fig. 6. F-score, Recall and Precision change curve with prediction threshold.

2.5   Model Ensemble

The ensemble of sub-models can effectively improve the generalization ability of the prediction model. We finally selected 6 sub-models that perform well on the validation set. These sub-models use XGBoost as the basic algorithm. The difference between them is mainly in the preprocessing, such as different SMART features, different feature extraction methods and sliding windows. Detailed parameters of these six sub-models are shown in the Fig. 7. The final prediction probabilities of the integrated models are obtained by averaging the prediction probabilities of these six models.

Model	SMART	Feature extraction	Sliding window	Positive samples	Sampling positions
Model01	5_raw, 187_raw, 197_raw, 198_raw, 199_raw	std	3, 5, 7	7	0, 1, 2, 3, 4, 5, 6
Model02	5_raw, 187_raw, 192_raw 197_raw, 198_raw, 199_raw	change rate change time	3, 5, 7	3	0, 1, 2 / 30, 60
Model03	5_raw, 187_raw, 193_raw 197_raw, 198_raw, 199_raw	change rate change time	3, 5, 7	5	0, 1, 2, 3, 4 / 30, 40, 50, 60
Model04	5_raw, 187_raw, 192_raw 197_raw, 198_raw, 199_raw	change rate change time	3, 5, 7	3	0, 1, 2, 30, 40
Model05	5_raw, 187_raw, 197_raw 199_raw	std	3, 5	7	0, 1, 2, 3, 4, 5, 6
Model06	5_raw, 187_raw, 192_raw 197_raw, 198_raw, 199_raw	std change rate change time	3, 5, 7	4	0, 1, 2, 3 / 30, 40, 50

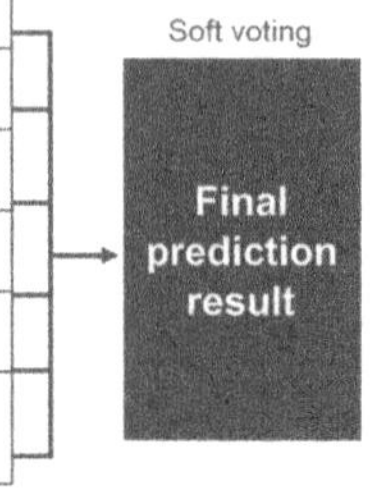

Soft voting

Final prediction result

Fig. 7. Model ensemble method.

The positive samples and sampling positions in the Fig. 7 are related to the sampling process in Sect. 2.3.

3   Evaluation

3.1   Evaluation Metric

According to the Alibaba's requirement, prediction engine predicts the failure disks in the next 30 days. We used the precision, recall and F-score evaluation metrics redefined in the competition rules [3].

Recall reflects the proportion of positive samples correctly judged to the total positive samples, and Precision reflects the proportion of true positive samples among the positive samples decided by the classifier. The higher Recall and Precision, the better. F1-Score is the weighted average of Recall and Precision. F-score takes into account both Recall and Precision.

The metrics are defined as follows:

$$Precision = \frac{n_{tpp}}{n_{pp}} \tag{3}$$

$$Recall = \frac{n_{tpr}}{n_{pr}} \tag{4}$$

$$F - score = 2 \times \frac{Precision \times Recall}{Precision \times Recall} \tag{5}$$

The following Table 2 explains n_{tpp}, n_{pp}, n_{tpr} and n_{pr}.

Table 2. Evaluation metric detail.

Number	Metric	Detail
1	n_{pp}	The number of disks that are predicted to be faulty in the following 30 days
2	n_{tpp}	The number of all the disks those truly fail among 30 days after the first predicting day
3	n_{pr}	The number of all the disk failures occurring in the k-day observation window
4	n_{tpr}	The number of truely faulty disks that are successfully predicted no more than 30 days in advance

3.2 Experimental Results

There are two steps in model verification stage. Firstly, we predict offline validation dataset, then select the optimal F-score value and corresponding prediction threshold. The prediction results of the offline validation dataset are shown in the Table 3. By ensembling several good sub-models together, the overall prediction performance can be improved. The best sub-model is Model_06, with F-score 34.21, while the integrated model's F-score reached 36.36, an increase of 2.15. Secondly, we predict online test dataset for final testing. The prediction precision was 52.42, the Recall was 32.31, and the F-score was 39.98.

Table 3. Experimental results offline.

Model	Precision	Recall	F1
Model01	37.6	29.7	33.21
Model02	40.6	27.2	32.58
Model03	47.4	23.4	31.36
Model04	36.4	30.4	33.10
Model05	36.3	31.0	33.45
Model06	35.6	32.9	34.21
ensemble	48.4	29.1	36.36

4   Summary

In large-scale data centers, disk is the component with the highest failure rate. Disk failure will seriously affect the stability and reliability of IT infrastructure. Based on the SMART data set of Alibaba Data Center, this paper designs and implements an efficient disk failure prediction system. The training process of the system consists of five parts: feature extraction, preprocessing, model training, model ensemble, and model verification. XGBoost algorithm is applied. After system-level optimization, the F-score achieves 39.98. In the competition jointly held by Alibaba and PAKDD, the effectiveness and versatility of our system was approved.

There are many viable ways of extending this work, such as: Applying transfer learning algorithm to solve the problem of insufficient samples of failed hard disks. Using ranking algorithms to make further improvements. Analyzing disks that are not reported in time or reported wrongly.

Acknowledgements. Thanks to Alibaba and PAKDD for organizing the PAKDD2020 Alibaba Intelligent Operation and Maintenance Algorithm Contest, which give us precious training data sets. This competition also gives us the opportunity to communicate with experts. We especially thank Inspur and the leaders for their support.

References

1. https://en.wikipedia.org/wiki/S.M.A.R.T
2. https://github.com/alibaba-edu/dcbrain/tree/master/diskdata
3. https://tianchi.aliyun.com/competition/entrance/231775/information
4. Hongzhang, Y., Yahui, Y., Yaofeng, T., et al.: Proactive fault tolerance based on "collection-prediction-migration-feedback" mechanism. J. Comput. Res. Dev. **57**(2), 306–317 (2020)
5. Sidi, L., Bing, L., Tirthak, P., et al.: Making disk failure predictions SMARTer! In: Proceedings of the 18th USENIX Conference on File and Storage Technologies (FAST 2020), Santa Clara, CA, USA, pp. 151–167 (2020)
6. Yong, X., Kaixin, S., Randolph, Y., et al: Improving service availability of cloud systems by predicting disk error. In: Proceedings of the 2018 USENIX Annual Technical Conference (USENIX ATC 2018), Boston, MA, USA, pp. 481–493 (2018)
7. Yanwen, X., Dan, F., Fang, W., et al: DFPE: explaining predictive models for disk failure prediction. In: 2019 35th Symposium on Mass Storage Systems and Technologies (MSST), Santa Clara, CA, USA (2019)
8. Ganguly, S., Consul, A., Khan, A., Bussone, B., Richards, J., Miguel, A.: A practical approach to hard disk failure prediction in cloud platforms: big data model for failure management in datacenters. In: 2016 IEEE Second International Conference on Big Data Computing Service and Applications, pp. 105–116. IEEE (2016)
9. Ma, A., et al.: RAIDShield: Characterizing, monitoring, and proactively protecting against disk failures. ACM Trans. Storage **11**(4), 17:1–17:28 (2015)
10. Nicolas, A., Samuel, J., Guillaume, G., Yohan, P., Eriza, F., Sophie, C.: Predictive models of hard drive failures based on operational data. In: 2017 16th IEEE International Conference on Machine Learning and Applications (ICMLA), pp. 619–625 (2017)
11. Tan, Y., Gu, X.: On predictability of system anomalies in real world. In: 2010 IEEE International Symposium on Modeling, Analysis & Simulation of Computer and Telecommunication Systems (MASCOTS), pp. 133–140. IEEE (2010)
12. Dean, D.J., Nguyen, H., Gu, X.: UBL: unsupervised behavior learning for predicting performance anomalies in virtualized cloud systems. In: Proceedings of the 9th International Conference on Autonomic Computing, pp. 191–200. ACM (2012)
13. Wang, Y., Miao, Q., Ma, E.W.M., Tsui, K.L., Pecht, M.G.: Online anomaly detection for hard disk drives based on mahalanobis distance. IEEE Trans. Reliab. **62**(1), 136–145 (2013)
14. http://docs.ceph.com/docs/master/mgr/diskprediction
15. Burges, C.J.C.: A tutorial on support vector machines for pattern recognition. Data Min. Knowl. Discov. **2**(2), 121–167 (1998). https://doi.org/10.1023/A:1009715923555

16. Chen, T, Guestrin, C: XGBoost: a scalable tree boosting system. In: Proceedings of the 22nd ACM SIGKDD International Conference on Knowledge Discovery and Data Mining, pp. 785–794. ACM (2016)

17. Liaw, A., Wiener, M., et al.: Classification and regression by randomForest. R News **2**(3), 18–22 (2002)

18. dos Santos Lima, F.D., Amaral, G.M.R., de Moura Leite, L.G., Gomes, J.P.P., de Castro Machado, J.: Predicting failures in hard drives with LSTM networks. In: Proceedings of the 2017 Brazilian Conference on Intelligent Systems (BRACIS), pp. 222–227. IEEE (2017)

19. Hochreiter, S., Schmidhuber, J.: Long short-term memory. Neural Comput. **9**(8), 1735–1780 (1997)

Summary of PAKDD CUP 2020: From Organizers' Perspective

Cheng He[1(✉)], Yi Liu[1], Tao Huang[1], Fan Xu[1], Jiongzhou Liu[1], Shujie Han[2], Patrick P. C. Lee[2], and Pinghui Wang[3]

[1] Alibaba Group, Hangzhou, China
hecheng.hc@alibaba-inc.com
[2] The Chinese University of Hong Kong, Hong Kong, China
[3] Xi'an Jiaotong University, Xi'an, China

Abstract. PAKDD2020 Alibaba AI Ops Competition is jointly organized by PAKDD2020, Alibaba Cloud, and Alibaba Tianchi platform. The task of the competition is to predict disk failures in large-scale cloud computing environments. We provide SMART (Self-Monitoring, Analysis and Reporting Technology) logs on a daily basis in the production environments without any preprocessing except anonymization. The SMART logs pose great challenges for analysis and disk failure prediction in real production environments, such as data noises and the extreme data imbalance property. In this paper, we first describe the competition task, practical challenges, and evaluation metrics. We then present the statistical results of the competition and summarize the key techniques adopted in the submitted solutions. Finally, we discuss the open issues and the choices of techniques regarding the online deployment.

Keywords: Disk failure prediction · Cloud computing · AI Ops

1 Introduction

In large-scale cloud computing environments, millions of hard disk drives are deployed to store and manage massive data [2]. With such large-scale modern data centers, disk failures are prevalent and account for the largest proportion among the hardware failures in cloud data centers [20]. Disk failures may lead to service performance degradation, service unavailability, or even data loss [5]. In order to provide cloud services with high availability and reliability, cloud service providers explore proactive fault tolerance approaches to predict disk failures in advance.

For more than a decade, researchers from academia and industry have made great progress in disk failure predictions. Recent studies [6,9,11,12,15–17,19,21, 22,24] conduct comprehensive analysis on SMART (Self-Monitoring, Analysis, and Reporting Technology) logs [4], and they also make use of machine learning

Jointly organized by PAKDD 2020, Alibaba Cloud and Alibaba Tianchi Platform.

algorithms to achieve accurate prediction results. However, as the public datasets are limited at scale in the whole community, it is difficult to apply state-of-the-art solutions directly into production environments due to the following more strict requirements for online deployment.

- **Prediction on a daily basis.** In the production environment, the monitoring system usually collects SMART logs on a daily basis. Thus, disk failure prediction is supposed to output prediction results at the same granularity. This requirement leads to more severe data imbalance problem.
- **Data noise.** Data noise is commonplace and attributed to labeling and sampling in the complicated production environments. Specifically, administrators often detect disk failures by using the self-defined expert rules, yet these rules may change over time and cannot cover the unknown disk failures, thereby leading to noise in labeling. Also, the data collection process may be interrupted by some incidents in production, which causes missing of data during data collection.

In this paper, we first describe the competition task and related challenges in Sect. 2. We then describe the details of our proposed evaluation metrics for disk failure prediction under requirements in production environments in Sect. 3. In Sect. 4, we highlight the mainstream and novel techniques adopted in the submitted solutions. We analyze the overall statistics of submitted results in Sect. 5. Finally, we discuss the open issues in practical deployment in Sect. 6.

2   Task Description

The task of PAKDD2020 Alibaba AI Ops Competition is about reliability and availability improvement in cloud computing environments, in particular, through the predictive maintenance of disk failures. The participants of the competition are required to predict disk failures in the future 30 days based on the historical SMART logs and failure tags. We provide the SMART logs of two hard disk models from the same manufacturer over the duration for more than one year. Each disk model contains more than 100 K independent hard disk drives. To the best of our knowledge, this is the largest public dataset by size for a single disk model for disk failure prediction. As this is a supervised learning task, in addition to SMART logs, we also provide labels contained in the failure tag file collected from our trouble tickets system. All the dataset and related descriptions are available at PAKDD Cup 2020 and Tianchi website: https://tianchi.aliyun.com/competition/entrance/231775/information?lang=en-us.

Table 1 shows the metadata of training and testing SMART logs of the disk models A1 and A2. The SMART logs contain 514 columns in total, including the disk serial number, manufacturer, disk model, data collecting time, and 510 columns of the SMART attributes. We denote the SMART attributes by "smart_n", where n is the ID of the SMART attribute. Each SMART attribute has a raw value and a normalized value, which are denoted by "smart_n_raw" and "smart_n_normalized", respectively. The SMART attributes are vendor-specific, while many of these attributes have a normalized value that ranges from 1 to 253 (with higher values representing better status). The initial default normalized value of the SMART attributes is 100 but can vary across manufacturers.

Table 2 shows the metadata of failure tags. It contains five columns including the disk serial number, manufacturer, disk model, failure time, and sub-failure type. In particular, the trouble ticket system in Alibaba Cloud detects disk failures using expert rules and reports the failure time (denoted by "fault_time") and sub-failure type (denoted by "tag") when failures occur. For the sub-failure types, we anonymize the names and map them into the numbers ranging from one to five, while we set the sub-failure type as zero by default for healthy disks.

Table 1. Metadata of training and testing sets.

Column name	Type	Description
serial_number	string	disk serial number code
manufacturer	string	disk manufacturer code
model	string	disk model code
smart_n_normalized	integer	normalized value of SMART-n
smart_n_raw	integer	raw value of SMART-n
dt	string	data collecting time

Table 2. Metadata of failure tags.

Column name	Type	Description
serial_number	string	disk serial number code
manufacturer	string	disk manufacturer code
model	string	disk model code
fault_time	string	time of failure reported in the trouble ticket system
tag	integer	ID of sub-failure type, ranging from 0 to 5

The competition has three rounds: preliminaries, semi-finals, and finals. In preliminaries, we provide the dataset of two disk models from the same manufacturer including the SMART logs and failure tags. We set the training period from July 31, 2017 to July 31, 2018 and the testing period from August 1, 2018 to August 31, 2018. During preliminaries, we first open the training and testing sets of disk model A1 for the participants. The participants can leverage the dataset to select features, construct machine learning model, and optimize their machine learning models. We then open the training and testing sets of disk model A2 for participants five days before the preliminaries deadline, so as to test the generalization of their methodology designed for disk model A1 on the new testing set (disk model A2). We select the top 150 teams with the higher prediction accuracy on disk model A2 into semi-finals.

In semi-finals, participants have the same training sets of disk model A1 and A2 with those in preliminaries, but the testing dataset is not available offline. Instead, we merge the testing data of disk models A1 and A2 together from September 1, 2018 to September 30, 2018 and put the resulting dataset into online testing environments. We require participants to submit their prediction solutions packed in docker files to online testing environments for predicting disk failures in the testing dataset (including both disk models A1 and A2). We select the top 12 teams with higher prediction accuracy into finals.

In finals, we require the 12 teams to present their solutions, including the main idea, design of machine learning models, and workflow details (e.g., feature selection and construction as well as optimization methods). We invite experts from industry and academia as our committee to score their presentations in terms of reasonability, novelty, and completeness. The final scores are comprised of two parts, i.e., 70% for prediction results and 30% for presentation.

In this competition, we summarize the following challenges in our dataset for disk failure prediction:

- **Extremely imbalanced data.** The data imbalance problem is a well-known challenge in the machine learning community [13], meaning that classifiers tend to be more biased towards the majority class. In production, when we predict disk failure on a daily basis, the data imbalance becomes more severe and the imbalance ratio of failed disks to healthy disks is less than 0.003% in our dataset. Therefore, data imbalance is a critical issue that needs to be addressed in both the competition as well as production environments.
- **Data noise.** Data noise is commonplace and attributed to many reasons, such as network failures, software malfunction/upgrades, system or server crashes, data missing, or anomaly collected data events in monitoring systems. All these events bring noise into the dataset and compromise the expected patterns of the dataset. Data noise cannot be ignored in disk failure prediction, as it can impair the prediction accuracy. How to design and apply proper techniques in dealing with the data noise problem is also a key to improving the accuracy of disk failure prediction solutions.

We encourage participants to leverage prior studies and state-of-the-art techniques to obtain domain knowledge of the SMART logs and disks during competition period in addition to our dataset and specifications. We also build a testing environment for participants to evaluate and reproduce the submitted solution, so as to guarantee the correctness of the prediction outputs.

3   Evaluation

We evaluate the prediction results of participants' submitted solutions based on three accuracy metrics, including precision, recall, and F1-score, as defined below.

- **Precision for P-window.** We define the *precision* as the fraction of actual failed disks being predicted over all (correctly and falsely) predicted failed disks. As our objective is to evaluate whether a failed disk being predicted is an actual failure within 30 days, we define the *P-window* as a fixed-size *sliding* window starting from the first time that a disk is predicted as failure, and set the length of the P-window as 30 days. Let T denote the start date and $T + k - 1$ denote the end date of the testing period (k as 30 days in our competition). Note that the P-window may slide out of the testing period. Figure 1 illustrates how we count true positive and false positive results. If the actual failure happens within the P-window (e.g., the 1st and 4th rows), we regard the failed disk as a correctly predicted one; otherwise (e.g., the 2nd and 3rd rows), we regard the disk as a falsely predicted one.
- **Recall for R-window.** We next define the *recall* as the fraction of actual failed disks being predicted over all actual failed disks. We define the *R-window* as a fixed-size window (not sliding window) from the starting date to the end date of the testing period with the length of 30 days in our case (i.e., from T to $T + k - 1$, where k is 30 days). Figure 2 shows how we count false positive, false negative, and true positive results. If a failed disk being predicted is not failed within the R-window (the 1st and 2nd rows), we regard the disk as a falsely predicted one; otherwise, we regard the failed disk as a correctly predicted one (the 4th and 5th rows). If an actual failed disk within the R-window is not predicted, we regard the failed disk as a missed one (i.e., false negative in the 3rd row).
- **F1-score.** We follow the classical definition of F1-score as $\frac{2 \times precision \times recall}{precision + recall}$. For easy comparison, we use F1-score as the participants' score in preliminaries and semi-finals.

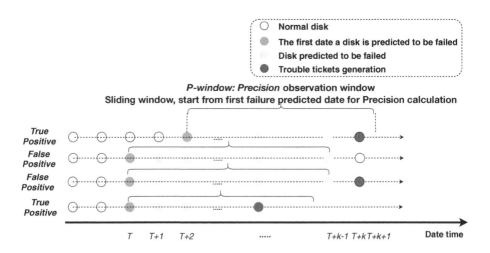

Fig. 1. Illustration of P-window.

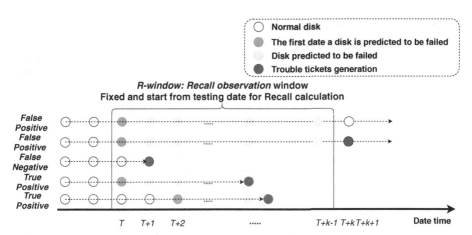

Fig. 2. Illustration of R-window.

In preliminaries, participants are required to submit prediction results in CSV format with four deterministic columns, including the manufacturer, disk model, disk serial number, and failure predicted date, based on our provided dataset. If one disk is predicted to be failed multiple times, we only take the earliest prediction date in the evaluation process and ignore the later ones. In semi-finals, we put the testing dataset on the cloud testing environments, so participants must submit their solutions packed with a docker image to predict disk failures on a daily basis. Then the auto-evaluation process gives a final score based on the aforementioned metrics. In finals, the top-ranking teams in the leaderboard are asked to present the strengths and weaknesses of their solutions. The final scores are comprised of two parts, 70% of prediction results in semi-finals and 30% of presentation results in finals.

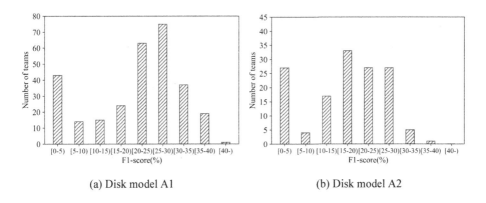

(a) Disk model A1

(b) Disk model A2

Fig. 3. F1-score distribution of prediction results of disk models A1 and A2 in preliminaries.

4   Statistics of Submissions

In this section, we analyze the statistics of submissions on the prediction accuracy distributions, team affiliation, and the time spent of participants on the competition.

4.1   Prediction Accuracy Distributions

We first analyze the prediction accuracy distributions in preliminaries. There are 1,173 teams registered in the competition and we received 3,309 valid submissions from 291 teams for failure prediction of disk model A1. Figure 3(a) shows the distributions of F1-score for predicting failures in disk model A1. From the figure, we can see that around 19.6% teams achieve the F1-scores higher than 30%, while nearly half (47.4%) of the teams' results fall into the interval between 20% and 30% F1-score. Figure 3(b) illustrates the distributions of F1-score for predicting failures in disk model A2. The top 141 teams uploaded their solutions 405 times in total. We notice that only 4.3% teams achieve the F1-scores higher than 30%, which is much worse than that for disk model A1. The reason may be mainly on the late opening of the training and testing data of disk model A2, so the teams only have 5 days for in-depth analysis. Most teams had to quickly transfer their knowledge, features, and even models learned from disk model A1 directly to disk model A2. It may lead to the severe overfitting problems.

In semi-finals, 76 teams submitted 3,299 valid solutions to predict disk failures from the mixed disk models A1 and A2. Figure 4 shows that more than 61.8% teams obtain the F1-scores higher than 30%, which is much better than the results in preliminaries (19.5% teams for disk model A1 and 4.3% teams for disk model A2). It also indicates that after passing preliminaries, participants

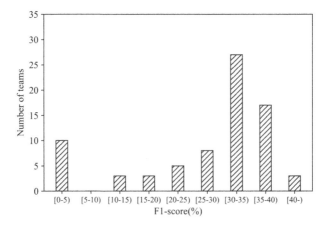

Fig. 4. F1-score distribution of prediction results of mixed disk A1 and A2 in semi-finals.

Table 3. Presentation results and semi-final scores of top teams.

Final ranking	Presentation score	Semi-final score
1	25.125	49.068
2	26.250	42.513
3	25.875	40.466
4	25.000	39.977
5	25.375	38.177
6	24.250	38.575
7	22.250	38.792
8	22.125	37.002
9	21.500	37.215
10	19.875	38.251
11	19.125	37.116

can pay more attention to feature engineering, modeling, and optimizing their solutions to improve their prediction results.

Finally, the top 12 teams entered the finals. Table 3 shows the presentation results and the semi-final score. The final score consists of 30% of presentation results and 70% of semi-final score. The presentation results in finals are evaluated in four major aspects, including novelty (10 points), reasonability (10 points), integrity (5 points), and presentation performance (5 points), by a committee of experts from academia and industry.

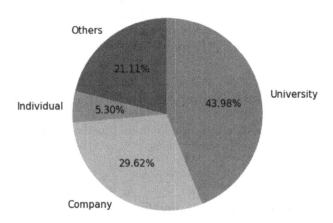

Fig. 5. Occupation analysis from registration information

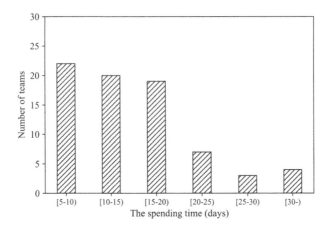

Fig. 6. Days spent to reach 85% of each team's best prediction results

4.2 Team Affiliation and Time Spent

We next present the team affiliation. Figure 5 shows that more than 70% of teams come from universities and companies. Most of them are familiar with machine learning and data mining, while very few of them have a strong background in storage reliability.

We also analyze the time spent of participants on the competition in order to know how much time participants need to apply their knowledge into a new field. Figure 6 shows the histogram results of days spent when reaching 85% of the best prediction accuracy for a team. We choose the 85%-mark as it can reflect that participants have completed the most part of their solutions. We notice that more than 81.3% of the teams can complete the majority of the task within 20 days, while around 29.3% of the teams spend less than 10 days on the task. These results provide us very useful insights for our future promotion of AI OPs to the community.

5 Mainstream Methodology and Highlighted Techniques

In this section, we summarize the techniques and methods applied in the competition based on the reviewed submissions.

5.1 Mainstream Methodology

From the reviewed submissions, the workflow of most solutions is comprised of four components, including data preprocessing, training sample generation, feature engineering, and modeling.

- **Data preprocessing.** As we describe before, there exists data noise in the SMART logs and failure labels. Thus, data preprocessing becomes an essential step that is applied by almost all teams. Most of the teams apply simple methods to solve the problem. For example, they drop the samples with missing data directly or interpolate missing data by forward filling or backward filling.

- **Training sample generation.** Re-sampling techniques [3] are popular approaches for dealing with data imbalance. Existing methods include *oversampling* (e.g., SMOTE [7]), which directly duplicates positive samples, and *undersampling* (e.g., cluster-based undersampling [23]), which selects a subset of negative samples randomly with a predefined ratio.
- **Feature engineering.** Feature construction and feature selection are two important steps for feature engineering. In the competition, almost all teams exploit sliding-window-based statistical features with various window lengths, such as difference, mean, variance, and exponentially weighted moving-average values. Some teams also select important features based on correlation analysis and remove the weakly correlated SMART attributes to failures.
- **Modeling.** Most teams formulate disk failure prediction as a binary classification problem and use tree-based ensemble models, such as random forests and the decision-tree variations. Among them, LightGBM [14] and Xgboost [8] are applied the most because of their efficiency in execution time and memory usage.

5.2   Highlighted Techniques and Novel Ideas

In addition to conventional approaches, we notice that participants also propose and try many novel ways in the competition. We highlight some approaches and categorize them into the aforementioned four components.

In data preprocessing, a team proposes the cubic spline interpolation method to solve data missing problem, and their experiment results show that it can improve the benchmark result of F1-score by more than 3%.

In training sample generation, two teams apply different methods from the above resampling techniques, i.e., GAN [10] and self-paced ensemble model [18]. GAN augments positive samples, while self-paced ensemble model is an undersampling method for downsampling negative samples. From the experimental results, these two methods become useful complements to the re-sampling techniques for mitigating the data imbalance problem.

In feature engineering, some teams propose different feature construction methods based on data analysis. They analyze the distance of failure occurrences, distributions of disk lifetime, and data missing ratio. We find that each of the methods, as well as the combinations of the methods, can improve the overall prediction results.

In modeling, in addition to using binary classification models, several teams formulate the problem as a multi-label classification or a regression problem, which can result in a higher F1-score. Furthermore, a team designs a two-layer stacking model, in which the second layer uses different features from the first layer and aims to reduce the number of false positives. All the highlighted creative methods give us more inspirations and will be helpful for all of us in future exploration.

6 Discussion

From the competition perspective, we admit that some tricks are very useful for improving prediction results. However, from an online deployment standpoint, besides the balance of solution complexity and online performance, we pay more attention to the interpretation of the model and results. Thus, some features and methods applied in this competition are debatable for production deployment. We list some of them for open discussion:

- *Should we take SMART-9 (power-on hours) as one of the important features?* The raw value of SMART-9 is a cumulative value and indicates the lifetime of a disk. This value should increase by 24 h for normal disks if we collect the data on a daily basis. Also, we do not find any evidence in the production environment that the statistical features based on SMART-9 have a significant correlation with disk failures.
- *Should we do bitwise decoding for the SMART attributes, such as SMART-1 (read error rate), SMART-7 (seek error rate), SMART-188 (command time-out), and SMART-240 (head flying hours)?* The raw values of these SMART attributes are vendor-specific and are often meaningless as decimal numbers [4]. Some teams construct statistical features based on these SMART attributes without bitwise decoding, but we are still unsure whether this method is reasonable and effective.
- *How should we interpret tree-based ensemble models?* Tree-based ensemble models, like random forest and GBDT, are popular and widely used in the competitions. However, these models can only provide overall feature importances without clear information on the individual predicted output to support engineers for locating and solving disk failures.

Another interesting phenomenon is the limited usage of cutting-edge techniques like deep learning. One possible reason is that deep learning is a kind of data-hungry methodology. Even though we have opened the large-scale datasets, the data size is still insufficient for teams to build sophisticated deep learning neural networks.

Besides the techniques and methods mentioned and applied in this competition, we have published part of our progress in [11,12]. Although the results cannot be directly comparable with this competition because of the differences of datasets, techniques like data preprocessing (correlation analysis, spline interpolation, automated pre-failure backtracking, denoising etc.), feature engineering, and modeling are fully tested and applied in production environments.

In the future, we will keep pushing this area forward by gradually opening more anonymized datasets from different disk models, manufacturers, performance data, and system logs in addition to the SMART logs and failure tags. All datasets will be made available on the official Github website [1]. Also, other AI OPs tasks, such as memory error prediction, server downtime prediction, server cluster auto-healing, application-level intelligent operations, will also be taken into consideration. With more data and information, we encourage the community to find more interesting and challenging problems in this field for further research and breakthrough.

Acknowledgement. We would like to express our gratitude to the PAKDD conference, especially Prof. Mengling Feng and Prof. Hady Wirawan Lauw for the great help and coordination. We thank Jingyi Huang, Ting Wang and Lele Sheng from the Tianchi platform for their hard work on test environment building, competition organization and coordination.

References

1. Alibaba Cloud AIOPs open datasets. https://github.com/alibaba-edu/dcbrain/tree/master/diskdata
2. Data Age 2025 - The Digitization of the World From Edge to Data. https://www.seagate.com/our-story/data-age-2025/
3. Resampling: Oversampling and Undersampling. https://en.wikipedia.org/wiki/Oversampling_and_undersampling_in_data_analysis
4. Wiki on S.M.A.R.T. https://en.wikipedia.org/wiki/S.M.A.R.T
5. Alagappan, R., Ganesan, A., Patel, Y., Arpaci-Dusseau, A.C., Arpaci-Dusseau, R.H.: Correlated crash vulnerabilities. In: Proceedings of USENIX OSDI (2016)
6. Botezatu, M.M., Giurgiu, I., Bogojeska, J., Wiesmann, D.: Predicting disk replacement towards reliable data centers. In: Proceedings of ACM SIGKDD (2016)
7. Chawla, N.V., Bowyer, K.W., Hall, L.O., Kegelmeyer, W.P.: SMOTE: synthetic minority over-sampling technique. J. Artif. Intell. Res. **16**, 321–357 (2002)
8. Chen, T., Guestrin, C.: XGBoost: a scalable tree boosting system. In Proceedings of ACM SIGKDD (2016)
9. Eckart, B., Chen, X., He, X., Scott, S.L.: Failure prediction models for proactive fault tolerance within storage systems. In: Proceedings of IEEE MASCOTS (2009)
10. Goodfellow, I., et al.: Generative adversarial nets. In: Proceedings of NIPS (2014)
11. Han, S., Lee, P.P.C., Shen, Z., He, C., Liu, Y., Huang, T.: Toward adaptive disk failure prediction via stream mining. In: Proceedings of IEEE ICDCS (2020)
12. Han, S.: Robust data preprocessing for machine-learning-based disk failure prediction in cloud production environments. arXiv preprint arXiv:1912.09722 (2019)
13. Japkowicz, N.: The class imbalance problem: significance and strategies. In: Proceedings of ICAI (2000)
14. Ke, G., et al.: LightGBM: a highly efficient gradient boosting decision tree. In: Proceedings of NIPS (2017)
15. Li, J., et al.: Hard drive failure prediction using classification and regression trees. In: Proceedings of IEEE/IFIP DSN (2014)
16. Li, J., Stones, R.J., Wang, G., Li, Z., Liu, X., Xiao, K.: Being accurate is not enough: new metrics for disk failure prediction. In: Proceedings of IEEE SRDS (2016)
17. Li, P., Li, J., Stones, R.J., Wang, G., Li, Z., Liu, X.: ProCode: a proactive erasure coding scheme for cloud storage systems. In: Proceedings of IEEE SRDS (2016)
18. Liu, Z., et al.: Self-paced ensemble for highly imbalanced massive data classification. arXiv preprint arXiv:1909.03500 (2019)
19. Lu, S.L., Luo, B., Patel, T., Yao, Y., Tiwari, D., Shi, W.: Making disk failure predictions SMARTer! In: Proceedings of USENIX FAST (2020)

20. Wang, G., Zhang, L., Xu, W.: What can we learn from four years of data center hardware failures? In: Proceedings of IEEE/IFIP DSN (2017)
21. Xiao, J., Xiong, Z. , Wu, S. , Yi, Y., Jin, H., Hu, K.: Disk failure prediction in data centers via online learning. In: Proceedings of ACM ICPP (2018)
22. Xu, Y., et al.: Improving service availability of cloud systems by predicting disk error. In: Proceedings of USENIX ATC (2018)
23. Yen, S.-J., Lee, Y.-S.: Cluster-based under-sampling approaches for imbalanced data distributions. Expert Syst. Appl. 36(3), 5718–5727 (2009)
24. Zhu, B., Wang, G., Liu, X., Hu, D., Lin, S., Ma, J.: Proactive drive failure prediction for large scale storage systems. In: Proceedings of IEEE MSST (2013)

Author Index

Printed in the United States
By Bookmasters